BIOCHEMISTRY RESEARCH TRENDS

GLUTAMIC ACID

CHEMISTRY, FOOD SOURCES AND HEALTH BENEFITS

Biochemistry Research Trends

Additional books in this series can be found on Nova's website under the Series tab.

Additional e-books in this series can be found on Nova's website under the e-book tab.

Protein Science and Engineering

Additional books in this series can be found on Nova's website under the Series tab.

Additional e-books in this series can be found on Nova's website under the e-book tab.

Biochemistry Research Trends

Glutamic Acid

Chemistry, Food Sources and Health Benefits

Dantel M. J. Balcazar
and
Esmeralda A. Reinoso Perez
Editors

New York

For permission to use material from this book please contact us:
Telephone 631-231-7269; Fax 631-231-8175
Web Site: http://www.novapublishers.com

NOTICE TO THE READER

Library of Congress Cataloging-in-Publication Data

Glutamic acid : chemistry, food sources and health benefits / editors, Dantel M. J. Balcazar and Esmeralda A. Reinoso Perez.
p. cm.
Includes bibliographical references and index.
ISBN 978-1-62257-236-6 (hardcover)
1. Glutamic acid--Metabolism. 2. Glutamic acid--Physiological effect. I. Balcazar, Dantel M. J. II. Reinoso Perez, Esmeralda A.
QP562.G5G58 2012
572'.65--dc23
2012021375

Published by Nova Science Publishers, Inc. † New York

Contents

Preface

Poly (y-glutamic acid), a novel polyanionic and multifunctional macromolecule synthesized by Bacillus species, has attracted considerable attention because of its eco-friendly, biodegradable and biocompatible characteristics. Recently, its application in a wide range of fields such as food, agriculture, medicine, hygiene, cosmetics and the environment has been explored. This book discusses the chemistry, food sources and health benefits of glutamic acid. Topics include the self-organization and stability of poly a-glutamic acid with cationic surfactants; metabolism and physiology of glutamate in chickens; different roles and applications of glutamic acid in the biomedical field; glutamic acid in food and its thermal degradation in acidic medium; application of a natural biopolymer poly (y-glutamic acid) as a bioflocculant and adsorbent for cationic dyes and chemical mutagens; and biodegradability of the y- and a-enantiomeric forms of the poly (glutamic acid) by bacillus licheniformis NCIMB 11709.

Chapter 1 - *Poly-Glutamic Acid (PGA)* is a biopolymer consisting of repeated units of amino acid *Glutamic Acid (GA),* in a certain combination of D- and L-forms. A striking feature of GA is that it bears two carboxyl groups, compared to the neutral aminoacids which bear just one. This second carboxyl group -COOH, which in the customary sequence of aminoacids in proteins is considered as an acidic side-chain, provides extra possibilities for binding either to amino-groups $-NH_2$ of other GA monomers, e.g.[$-COOH + -NH_2 \rightarrow$ -CO-NH- $+H_2O$] leading thus to the celebrated γ- Polyglutamic acid; or to various metal cations M^{+n} , e.g. [$n(-COOH) + M^{+n} \rightarrow n(-COO)-M + nH^{+}$] leading to chelation compounds, or finally to other cationic species like surfactants $CTA^{+} B^{-}$, e.g.[$-COOH + CTA^{+} B^{-} \rightarrow$ -COO-CTA +HB] leading to *micelles* which can be used as precursors for the development of mesoporous materials.There is also the possibility for a simultaneous occurrence of the three above reactions, leading to a rich hybridic chemistry of this compound. Besides of the above substitution reactions, the GA and the PGA bear an asymmetric carbon atom, termed *α-carbon,* which results in the *chiral* forms of GA, namely the right-hand isomer, or D-Glutamic Acid and the left-hand isomer, or L-Glutamic Acid. This handness can be also expressed in the polymeric forms leading to Poly-(D) Glutamic Acid and Poly-(L) Glutamic Acid respectively. Such chiral polymeric forms of PGA, combined with cationic surfactants, lead to chiral micelles. Such chiral micelles show subtle but insisting indications of differentiated resistance to hydrolysis a fact which may be related to the well known enigmatic persistence—but still unexplained— of the L-aminoacids and the right-hand *α-helix* of polyaminoacids in the living world. This review highlights recent developments in the chemistry of PGA in relation to its self-organization with cationic surfactants CTAB, the structure and applications of the resulting micelles, and their distinct stability in aquatic environment which may be related to the different forms of ortho- and para- water in nature.

Chapter 2 - Glutamate is one of the proteinogenic and nutritionally non-essentional amino acids, and is precursor of glutamine, arginine, proline and glutathione. Glutamate also serves as the precursor for the synthesis of the inhibitory γ-aminobutyric acid (GABA) in GABAergic neurons. Among these metabolites, synthesis of arginine and proline in chickens is different from the mammals. Chickens lack carbamyl phosphate synthetase I and have relatively little ornithine transcarbamylase activities. Chickens are also unable to synthesize ornithine from glutamate and proline. These characteristics make the absolute dietary requirement of arginine in chickens. In addition, the biosynthesis of proline from glutamate is not rapid and not supports maximum growth in chicks. On the other hand, glutamate, an excitatory amino acid, acts as a neurotransmitter in the central nervous system. It can induce neuronal activity with powerful stimulatory effects. However, the central administration of glutamate attenuates stress-induced behaviors and triggers sleep-like behavior in neonatal chicks. Furthermore, glutamate plays major roles in the development of normal synaptic connections in the brain. Glutamate concentration increased in the brain with the advancement of age in developing chicks. In conclusion, glutamate is an important metabolite which plays significant roles in the development and maintenance of peripheral and central tissues in the chicken.

Chapter 3 - Poly(glutamic acid) (PGA) is a carboxylic functionalized polypeptide that has been object of intensive research.

PGA is a hydrophilic, biodegradable, and naturally available biopolymer usually produced by various strains of Bacillus. Its biological properties such as nontoxicity, biocompatibility, and nonimmunogenicity qualify it as an important biomaterial that can be easily assimilated *in vivo* rendering it usable in the field of medicine, pharmaceutics, cosmetics, food industry and others.

Poly (glutamic acid) polymer possesses appropriate physicochemical properties that allow it to be used as a reducing agent, capping agent or for the development of polymer–drug conjugates in order to achieve selective and targeted delivery of drugs.

This review article deals with the synthesis, physiochemical properties, and various biomedical applications of poly (glutamic acid).

Chapter 4 - Glutamic acid is an amino acid naturally occurring in many foods and it is responsible for umami taste. For this reason, it has been widely used as a food additive and flavor enhancer as monosodium glutamate.

Glutamic acid is stable under standard conditions. However, high temperature or extreme pH conditions may induce racemization. In addition, it is a good substrate for non-enzymic browning reactions (Maillard reaction).

A case study of the glutamic acid thermal degradation at acidic pH was carried out. Different grape musts were subjected to heating at 90 °C for 30 h by means of a lab-scale equipment emulating a real process. Model solutions were utilized to gain a deeper comprehension of this phenomenon and to explain the glutamic acid degradation pattern. Results showed that glutamic acid underwent degradation during grape must cooking, following the same trend observed in the model solutions. The amino acid was almost linearly reduced throughout all the cooking procedure yielding pyroglutamic acid as the main product of degradation.

Chapter 5 - Poly (γ-glutamic acid) (γ-PGA), a novel polyanionic and multifunctional macromolecule synthesized by Bacillus species, has attracted considerable attention because of its eco-friendly, biodegradable and biocompatible characteristics. Recently, its application in a wide range of fields such as food, agriculture, medicine, hygiene, cosmetics and environment has been explored. This book chapter reviews the literature reports on the application of γ-PGA as a flocculating agent, and adsorbent for cationic dyes and chemical mutagens, affected by several process parameters including pH, temperature, contact time, metal cations, concentration and molecular weight of γ-PGA.

Chapter 6 - Poly-γ-glutamic acid (γ-PGA) is a water soluble, polyanionic, extracellular polymer produced by several members of the genus *Bacillus*, the most notable among them *Bacillus subtillis* and *Bacillus licheniformis*. This polymer consists of D- and L-glutamic acid repeating units that are linked between the α-amino and γ-carboxylic acid functionalities. A peculiar characteristic of γ-PGA formed by *B. licheniformis* is that the polymer stereochemistry can be controlled by the composition of the medium used in the polymer production in the laboratory. γ-PGA is water soluble, biodegradable, edible and non-toxic to humans and the environment. Therefore, the potential applications of this polymer and its derivatives have been of growing interest in the past few years in a broad range of industrial fields such as food, cosmetics, medicine and water treatment.

Most of the reports published on the microbial or enzymatic biodegradation of microbial γ-PGA are only concerned with γ-D-PGA. In the present work, γ-L-PGA was biosynthesized using *B. licheniformis* NCIMB 11709, and the biodegradation of the different enantiomeric forms of PGA, as well as the α and γ forms (γ-L-PGA, γ-D-PGA, γ-D, L-PGA, α-L-PGA and α-D-PGA) of this bacterial strain, were studied. γ-PGA with L-units content of 70% was produced with a Mw of 1 million g/mole. *B. licheniformis* NCIMB 11709 did not grow when the α-forms of the polymer were used as sole carbon source, independently of the D or L contents. However, a decrease in the molecular weight of the α-L-PGA in the bacterial culture was observed, which was not due to hydrolytic degradation. On the other hand, this strain could degrade and use the polymer when it was in the γ-form, regardless of the D or L contents.

This finding can have an implication in defining bacterial factors involved in the degradation of the polymer. Moreover, the generated knowledge of the degradation of the different forms of the poly (glutamic acid) will facilitate selection of the better form according to the intended application.

In: Glutamic Acid
Editors: D. Balcazar and E. Perez
ISBN: 978-1-62257-236-6

Chapter 1

Self-Organization and Stability of Poly-α -Glutamic Acid (α-PGA) with Cationic Surfactants

Eleftheria K. Kodona, Eugenia Panou-Pomonis and Philippos J. Pomonis*
Department of Chemistry, University of Ioannina, Greece

Abstract

Poly-Glutamic Acid (PGA) is a biopolymer consisting of repeated units of amino acid *Glutamic Acid (GA),* in a certain combination of D- and L-forms. A striking feature of GA is that it bears two carboxyl groups, compared to the neutral aminoacids which bear just one. This second carboxyl group -COOH, which in the customary sequence of aminoacids in proteins is considered as an acidic side-chain, provides extra possibilities for binding either to amino-groups $-NH_2$ of other GA monomers, e.g.[-COOH + $-NH_2$ → -CO-NH- $+H_2O$] leading thus to the celebrated γ- Polyglutamic acid [1]; or to various metal cations M^{+n}, e.g. [n(-COOH) + M^{+n} → n(-COO)-M + nH^+] leading to chelation compounds, or finally to other cationic species like surfactants $CTA^+ B^-$, e.g.[-COOH + $CTA^+ B^-$ → -COO-CTA +HB] leading to *micelles* which can be used as precursors for the development of mesoporous materials

* E-mail: ppomonis@cc.uoi.gr.

[2].There is also the possibility for a simultaneous occurrence of the three above reactions, leading to a rich hybridic chemistry of this compound. Besides of the above substitution reactions, the GA and the PGA bear an asymmetric carbon atom, termed *α-carbon,* which results in the *chiral* forms of GA, namely the right-hand isomer, or D-Glutamic Acid and the left-hand isomer, or L-Glutamic Acid. This handness can be also expressed in the polymeric forms leading to Poly-(D) Glutamic Acid and Poly-(L) Glutamic Acid respectively. Such chiral polymeric forms of PGA, combined with cationic surfactants, lead to chiral micelles [3]. Such chiral micelles show subtle but insisting indications of differentiated resistance to hydrolysis [3, 4] a fact which may be related to the well known enigmatic persistence—but still unexplained— of the L-aminoacids and the right-hand *α-helix* of polyaminoacids in the living world. This review highlights recent developments in the chemistry of PGA in relation to its self-organization with cationic surfactants CTAB, the structure and applications of the resulting micelles, and their distinct stability in aquatic environment which may be related to the different forms of ortho- and para- water in nature.

1. The Glutamic Acid (GA) and its Position in the Galaxy of Aminoacids

Glutamic acid (GA) (M.W. = 147) belongs to the chemical category of 20 most common amino acids $H_2N\text{-}C^*(R)H\text{-}COOH$ that are the basic structural units of much of the functional material in the living world, i.e. the proteins, (Figure 1). The central carbon atom C^* in each aminoacid possesses *optical asymmetry* and is the reason that all these molecules may exist in two *chiral forms*, the left hand, or *L-isomer*, and the right hand or *D-isomer*. For some mysterious reasons all the natural aminoacids are the *L-isomers*. In the lab the two isomers are usually developed in equimolecular amounts without detectable preference to one of the optical isomers. Five out of those 20 common aminoacids contain ionizable side chains *(R)* that are of crucial importance to the mechanism of action of enzymes and to the binding of proteins to the charged sites of other small and large molecules or ions: Namely Lysine, Arginine and Hisidine contain amine-end side chains ($-NH_2$ or $-NH_3^+$) while Aspartic acid and Glutamic Acid bear carboxyl-end side chains ($-COOH$ or $-COO^-$). Glutamic acid is one of the more common *acidic amino acids*. The central chiral carbon atom (C*) in the aminoacids is termed *"alpha carbon" (α-C)*. If there are side chains, then the carbon atoms of the methylene groups are consequently termed “beta carbon” (β-C), “gama carbon” (γ-C), “delta carbon” (δ- C) and “epsilon carbon” (ε-C).

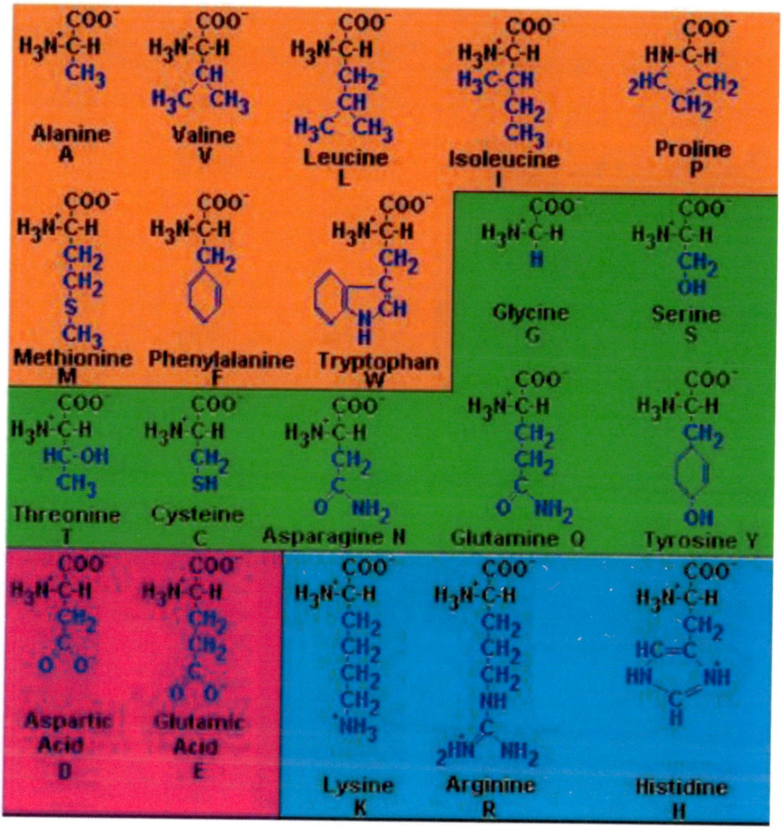

Figure 1. Each of the 20 most common aminoacids contain an "amine" group ($-NH_2$ or $-NH_3^+$) and a "carboxy" group (-COOH or $-COO^-$) (black in the diagram). The aminoacids vary in their side chains (blue in the diagram).The eight amino acids in the orange area are nonpolar and hydrophobic. The other aminoacids are polar and hydrophilic. The three amino acids in the light blue box are basic bearing an "amine" group in the side chain. The last two aminoacids in the magenta box are acidic bearing "carboxy" group in the side chain. Glutamic acid is a such acidic aminoacid. (Courtesy of Wikipedia).

Thus GA has the additional carboxyl side group in a γ-Carbon while lysine has the extra side amino group in a ε-Carbon. The irreplaceable property and function of aminoacids in nature is of course their ability to form *peptides* by combination of the acidic carboxyl terminal of one such molecule to the basic amino terminal of another (similar or different) molecule as shown in

Figure 2 for a dipeptide. The proteins are polypeptides where the building sequence of aminoacids in their chain is genetically controlled via the RNA and DNA.

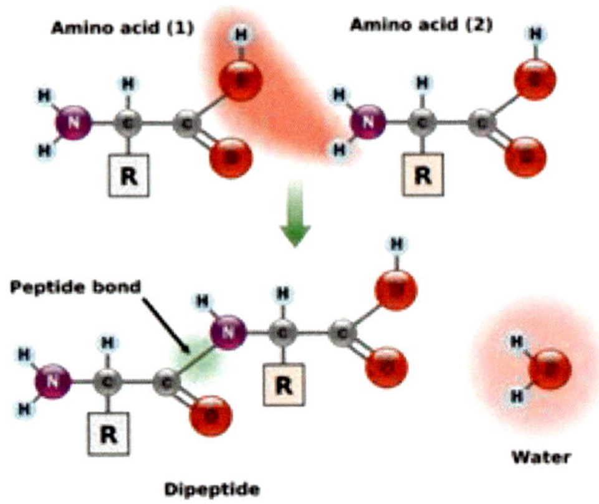

Figure 2. The peptide bond.

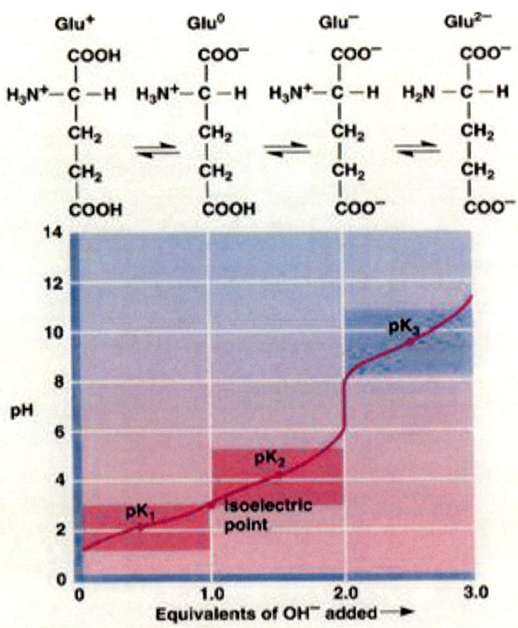

Figure 3. The three pK values for Glutamic acid are pK_1 =2.10 for the α-carboxylic group, pK_2 = 4.07 for the side carboxylic group and pK_3 = 9.47 for the α-amino group. The isoelectric point where the molecule is zero-charged corresponds to pH=3.0.

Now, if an aminoacid contains just one acidic and one basic group then its only possibility is to form *peptide bonds*. But if the molecule contains *additional ionizable side chains,* either basic (like Lysine, Arginine and Hisidine) or acidic (like Aspartic acid and Glutamic Acid), then such aminoacids possess the ability to form additional structures via that side group. Therefore those five aminoacids bearing charged side groups offer the possibility for additional (non peptide) chemistry. Glutamic acid is such a case since it is able to form bonds either via the "normal" or/and the "side" carboxyl group, leading to a variety of different products.

The ionization state of the basic ($-NH_2$) and/or acidic ($-COOH$) group, which is directly related to the pH values of the solution, determines the net charge of the aminoacids and is very important for their binding abilities to other acidic and/or basic sites respectively. For Glutamic acid the three pK values are shown in Figure 3 and are as follows: *pK for α-carboxylic acid =2.10, pK for side carboxylic acid = 4.07, pK for α-amino group = 9.47.*

2. The Poly-Glutamic Acid (PGA) and Its Hybridic Variances

Glutamic Acid possesses three functional groups and is optically asymmetric. This rare combination of chemical structure and properties is the reason for the rich chemistry of GA expressed in its polymeric forms and its hybridic compounds with either inorganic cations or surfactants. In addition all those compounds, polymers and/or hybridic materials, may be either the left (L) or the right (D) hand isomers. Let us be more specific as to the chemical structures obtained from GA.

(i) First, GA can form peptide bonds with other different aminoacids giving rise to various natural polypeptides and proteins. The skeletal bonds follow the sequence N- $^{\alpha}$C*-C-N- $^{\alpha}$C*-C-N- $^{\alpha}$C*-C-N- $^{\alpha C}$*-.... The carbons in the skeletal chain are all left-hand and the structure usually forms the celebrated right hand α-helix. Those compounds are the well known α-polypeptides (Figure 4).

(ii) Second, GA can form *peptide bonds with similar GA molecules* via the skeletal bond sequence N- $^{\alpha}C^{*}$-C-N- $^{\alpha}C^{*}$ C N- $^{\alpha}C^{*}$-C-N- $^{\alpha}C^{*}$- giving rise to artificial *α-Poly-Glutamic Acid (α-PGA)* (Figure 5). The only carbon atom participating in the skeletal sequence is the α-carbon. These compounds may be also chiral depending on the chirality of the original GA molecule and can be either *D-α-Poly-Glutamic acid or L-α-Poly-Glutamic acid.*

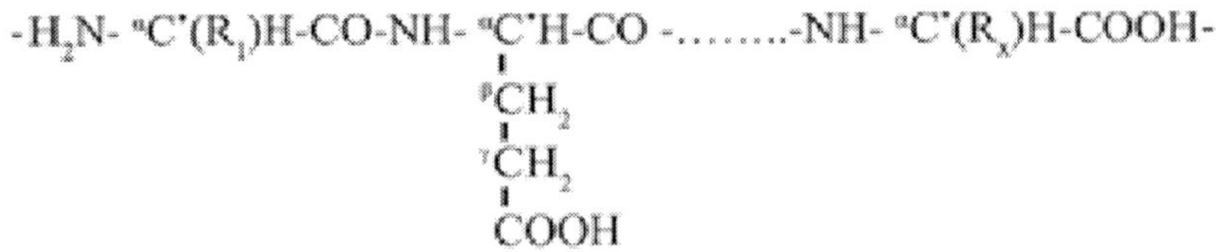

Figure 4. A natural polypeptide chain containing a single Glutamic Acid. The skeletal bond sequence is N- $^{\alpha}C^{*}$-C-N- $^{\alpha}C^{*}$-C-N- $^{\alpha}C^{*}$-C-N- $^{\alpha}C^{*}$- .... and forms an α-helix. The carbons $^{\alpha}C^{*}$ in the skeletal chain are all left-hand optically asymmetric. The side acidic chain is – $^{\beta}CH_2$ – $^{\gamma}CH_2$ – COOH. The proton H^+ of the carboxyl-group can suffer substitution reaction with another inorganic (for example Ca^{++}) or organic (for example R_4N^+) cation.

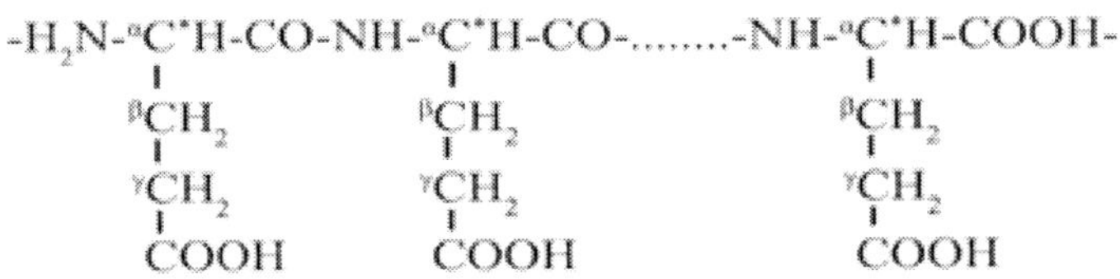

Figure 5. An artificial α-Poly-Glutamic Acid (α-PGA) chain. The skeletal bond sequence N- $^{\alpha}C^{*}$-C-N- $^{\alpha}C^{*}$-C-N- $^{\alpha}C^{*}$-C-N- $^{\alpha}C^{*}$- .... is similar to the natural polypeptides shown in Figure.4 . These compounds may be chiral of the form D-α-Poly-Glutamic acid or L-α-Poly-Glutamic acid depending on the chirality of the original GA molecule (D-α-GA or L-α-GA). The density of the acidic side chains is one every four (4) skeletal atoms and the protonic capacity quite high.

(iii) Third, GA can form *peptide - like bonds with similar GA molecules* via the side carboxyl. The resulting skeletal bond sequence is N- $^{\alpha}C^{*}$- $^{\beta}C$- $^{\gamma}C$-N- $^{\alpha}C^{*}$- $^{\beta}C$- $^{\gamma}C$ -N- $^{\alpha}C^{*}$- $^{\beta}C$- $^{\gamma}C$ -N- giving rise to *γ-Poly-Glutamic Acid (γ-PGA)* (Figure.6). The carbon atoms participating in the skeletal sequence are the α-Carbon, the β-Carbon and the γ-Carbon. γ-PGA is a naturally occurring biopolymer produced from Bacillus bacteria (Bacillus anthracis and Bacillus subtilis).

It is an essential constituent of Natto, used extensively as a health food in Japan and other Asian countries for hundreds of years. It is produced industrially by steaming soybeans and fermenting them by the Bacillus subtilis (natto).

It is also mentioned that γ-Poly-Glutamic Acid and ε-Poly- Lysine (see Figure 1) are, as far as we know, the only homo polypeptides detected so far in nature. The α-homo polypeptides, D-Poly-Glutamic acid and L-Poly-Glutamic acid, which will be discussed in this article, are not natural products but synthesized in the lab.

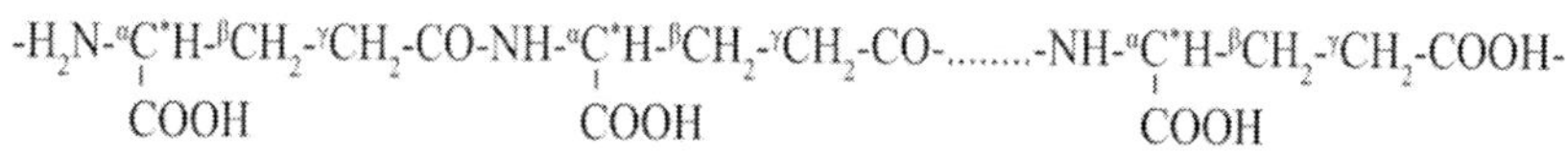

Figure 6. The γ-Poly-Glutamic Acid (γ-PGA) chain. The skeletal bond sequence is N- $^{\alpha}$C*- $^{\beta}$C- $^{\gamma}$C- N- $^{\alpha}$C*- $^{\beta}$C- $^{\gamma}$C -N- $^{\alpha}$C*- $^{\beta}$C- $^{\gamma}$C -N- ..... The density of the acidic side chains is one every six (6) skeletal atoms and the protonic capacity is lower compared to α-PGA chain.

(iv) Forth, the carboxyl groups in the side chains of *α-Poly-Glutamic Acid (α-PGA)* can react with *metal cations*, like for example Ca^{++} , Fe^{++} , Mg^{++} etc. (Figure 7) This chelating ability may be used in fertilizers for agriculture to facilitate the delivery of minerals to plants in order to correct mineral deficiencies and also to improve the absorption of minerals from supplements used in animal feed and human nutrition.

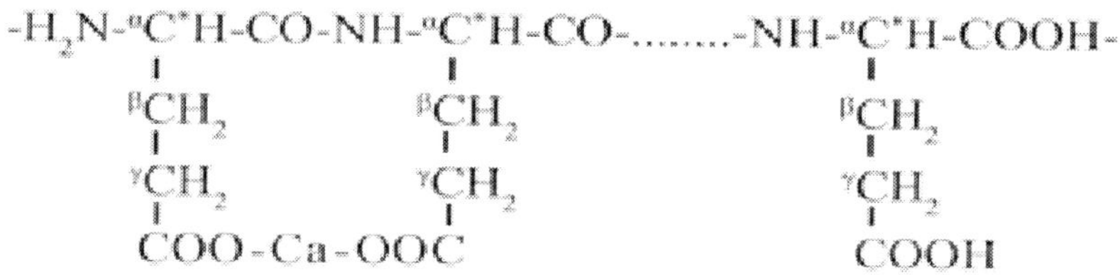

Figure 7. The proton H^{+} of the carboxyl-group in the side acidic chain – CH_2 – CH_2– COOH of α-PGA can suffer substitution reaction with another inorganic cation, for example Ca^{++} .

(v) Fifth, the carboxyl groups in the side chains of *α-Poly-Glutamic Acid (α-PGA)* can react with cationic surfactants $C_nH_{2n+1}N^{+}(CH_3)_3$ with $8 < n < 16$, termed usually *CTAB (Cetyl –Trimethyl -Ammonium Bromate)* via the simple reaction - COOH + CTAB → -COO-CTA+ HB (Figure.8).

The obtained hybridic compounds, under specific conditions, may form *micelles* of either the D- or the L-form which in turn can be used as precursors for the development of *silicate MCM-type materials*.

(vi) Sixth, the *optically isomer* (chiral) compounds, D-Poly-Glutamic Acid (D-PGA) or L-Poly-Glutamic Acid (L-PGA), or their corresponding hybridic forms D-PGA/CTAB and L-PGA/CTAB (Figure.5), can interact with the most common compound on earth, which is water, and suffer hydrolysis. Now it is well established, but perhaps less well known, that water in nature exists in two different *spin-isomer* forms, the *ortho water* (o-H_2O) and the *para-water* (p-H_2O) in a 3:1 ratio.

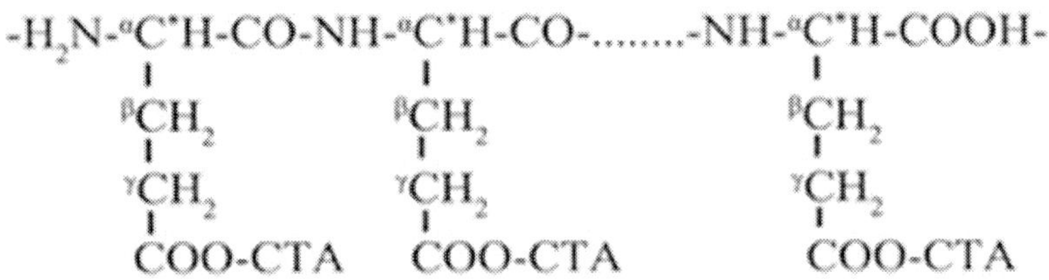

Figure 8. The proton H+ of the carboxyl-group in the side acidic chain of α-PGA can suffer substitution reaction with another organic cation, like the alkyl quaternary ammonium ions, $C_nH_{2n+1}N^+(CH_3)_3$ ($8 < n < 16$) $CTA^+ B^-$ (Cetyl –Trimethyl - Ammonium Bromate), - COOH + CTAB → -COO-CTA+ HB.

These two forms of water are discriminated by the orientation of the hydrogen atoms nuclear spins-either parallel (total spin I = 1 and nuclear statistical weight 2I + 1 = 3 for the paramagnetic o-H_2O) or antiparallel (total spin I = 0 and nuclear statistical weight 2I +1 = 1 for the nonmagnetic p-H_2O). The big problem is that those two forms of water it difficult to be isolated/separated and extremely difficult to be detected. Nevertheless there are indirect indications that they may affect differently the stability and the rate of hydrolysis of the chiral enantiomers- see Figure 9.

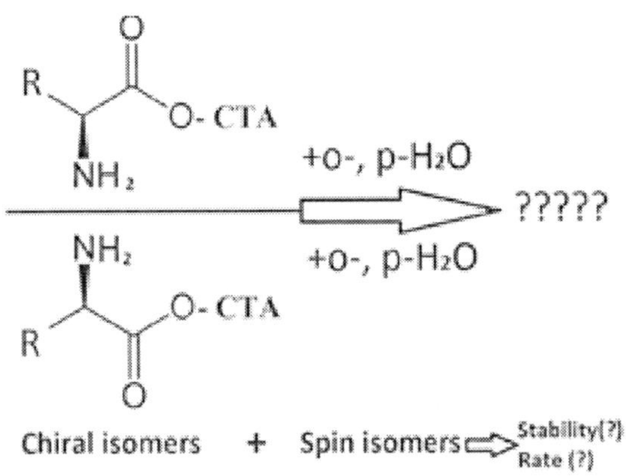

Figure 9. The hydrolysis of chiral isomers D-PGA/ CTAB and L-PGA/ CTAB may be influenced by the spin isomers ortho-water (o-H_2O) and para-water (p-H_2O).

In the present article we shall deal mainly with points (v) and (vi) but inevitably the discussion involves matters related to the other points (i, ii, iii and iv) as well.

3. Self-Organized Structures Obtained via Interaction of L-PGA with Cationic Surfactants CTAB

The α-PGA (Figure 5) is essentially a polyelectrolyte of biological origin. A similar non-biological polyelectrolyte is the well known and commonly used Poly-Acrylic Acid (PAC). In the middle 1980's it was shown that the interaction of such artificial (non-biological) polyelectrolytes with cationic surfactants, like the long-chain alkyl quaternary ammonium ions, $C_nH_{2n+1}N^+(CH_3)_3$ with $8 < n < 16$, (Figure 7) may lead to self-assembled micelles of various structures [5-7].

In the middle 1990's the group of Antonietti [8, 9] showed that the system PAC- C_nTAB forms mesophases with interesting polymeric properties. Consequently in the middle 2000's Pantazis et al. [10-12] recognized that the system PAC- C_nTAB can be employed as substrate for the development of M41S- type ordered mesoporous silicate materials, like MCM -41 [13-14], MCM-48 [15-16] and SBA [17-18], which were invented and became extremely popular in the 1990's.

The method of synthesis of M41S porous solids described in [10-12] is simple and based on the concept of *self-organization*: It takes place in one step via a simple titration with dilute NH_3 (0.1 N) of an acidified solution containing PAC, plus C_nTAB, plus TEOS (Tetra-Ethyl-Oxide Silicate, Si-$(OEt)_4$), as a source of silica, plus a metal nitrate salt, if doping with metal oxidic species like CuOx, CeOx, and CoOx etc, is wished.

The same system PAC - C_nTAB can be used for the development of either MCM-41 or SBA materials by simple alternation of surfactant C_{16}TAB to C_{14}TAB. Last but not least, the introduction of some cations like Mg and Sr (but not Ca!) as well as Co, in the system PAC-C_nTAB-TEOS results in morphogenesis of various decorated structures reminding biominerals [12, 19].

In Figure 10 the mechanism of self-assembly is shown between the three particular species PAC + C_nTAB + TEOS. Initially orderly packed micelles in hexagonal fashion are obtained. Then careful firing and removal of the internal organic content from the cylindrical micelles leads to MCM-41 silicate solids with ordered porosity [12].

The self-assembly is driven by the complexation of side acid groups of polyelectrolyte PAC with the cationic surfactant CTAB forming the backbone of the cylindrical micelle (reaction 1). The hydrophilic head $C_nH_{2n+1}\,N^+(CH_3)_3$ of CTA is located outwards and towards the aquatic phase where anionic oxy-hydroxy-silicate species from the hydrolysis of TEOS [e.g. $Si(OEt)_4 + H_2O \rightarrow Si(OH)_4 + 4EtOH$, $xSi(OH)_4 \rightarrow Si_x(OH)_yO_z^{-n} + kH_2O$) are formed and attracted/precipitated on the outer cylindrical surface (reaction 2).

Then careful firing and burning of the inner organic part leaves the hexagonically packed empty cylindrical shells - the well known MCM-41 structure. The hexagonal packing is certified by XRD diffractograms and the Bragg refraction (100) at low angles ($2 < 2\theta < 3$) as shown in the inset of Figure 10.

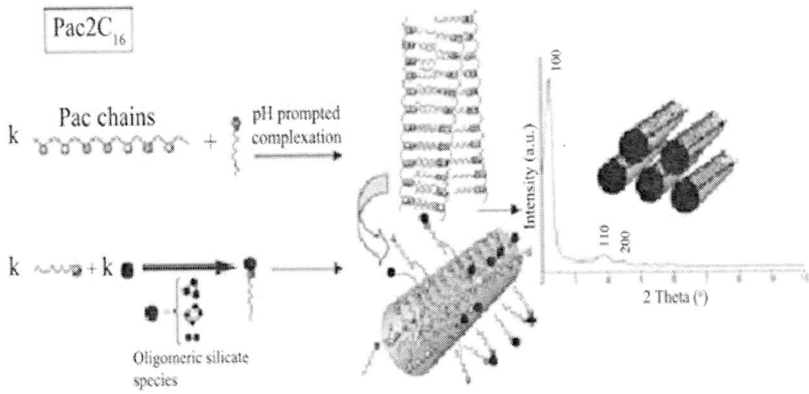

Figure 10. Proposed mechanism of self-assembly between Pac + C_nTAB + TEOS. Initially orderly packed micelles in hexagonal fashion are obtained. Then careful firing and removal of the internal organic content from the cylindrical micelles leads to MCM-41 silicate solids with ordered porosity [12]. The hexagonal packing is certified by XRD and the Bragg refraction (100) at low angles ($2<2\theta<3$) as shown in the inset. Reprinted with permission from [12].Copyright 2012 American Chemical Society.

The calculated d-space from the Bragg law $n\lambda = 2d\sin\theta$ corresponds to the distance between centers of the cylindrical pores.

Now, one of the most abundant natural acidic poly-electrolytes in nature, analogues to PAC, is the poly-aminoacids Poly-Aspartic Acid (PAA) and the Poly- Glutamic Acid (PGA).

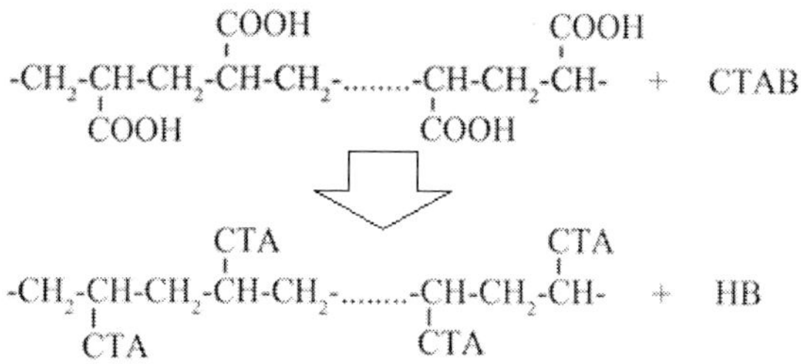

(1)

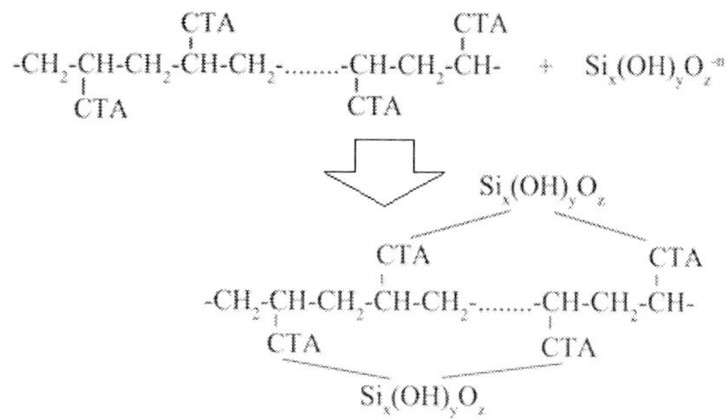

(2)

The abundance of those substances in the natural world has very specific reasons and targets: They act as binders of basic cations, mainly Ca (II), either for stabilizing the structure of shells in larcae and other protozoans which is made of calcite $CaCO_3$ crystallites [20], or for destabilizing the development of ice crystallites in the cold water fishes and other organisms [21-23] or either for modifying the development of hydroxyapatite crystallites in living organisms [24-26]. The successful development of M41S ordered mesoporous silicate materials using the system PAC-CTAB –TEOS [10-12, 19], triggered the idea to employ PGA as a blueprint molecule [2, 3], directing the development of the MCM-41 structures [13-18]. So, both micellar mesophases based on PGA-CTAB and silicate solids based on PGA-CTAB –TEOS were developed that are shown in Table 1.

The homo-polymer used was the L-α-PGA and the surfactants $C_{14}TAB$ and the $C_{16}TAB$. In some of the samples a small amount of EtOH (0.5%v) was added to assist dissolution/self-organization. Details about the method of synthesis and the characterization can be found in [2]. In the same Table 1 the Bragg refraction (100) at low angles is shown corresponding to the distance between the centers of cylindrical micelles (dried materials) or the centers of silica mesopores (calcined solids).

Table 1. Materials developed from the systems PGA-CTAB and PGA-CTAB –TEOS. Synthesis took place in one step via titration with dilute NH_3 (0.1 N) of an acidified solution containing PAG + C_nTAB +TEOS in ratio (–acidic site chain COOH:CTAB: Si) = 1:1:~5. The polymer used was the L-α-PGA. The pH of isolation was in the range 7-8 (see Figure 3) with no apparent differentiation in the d_{100}-space [2]

Sample	XRD structure	d_{100} (nm)	Observations
L-α-PGA/C_{14}TAB/7.5	Hexagonal	3.60	Organic hybridic micelles dried at 60° C
L-α-PGA/C_{14}TAB/8.5	Hexagonal	3.60	Organic hybridic micelles dried at 60° C
L-α-PGA/C_{14}TAB/TEOS/ uncalc	Hexagonal	4.13	Inorganic-Organic mesophases dried at 60° C
L-α-PGA/C_{14}TAB/TEOS/ calcin	Hexagonal	3.90	Silicate mesostructures calcined at 650° C
L-α-PGA/ C_{16}TAB/7.5	Hexagonal	4.10	Organic hybridic micelles dried at 60° C
L-α-PGA/ C_{16}TAB/8.5	Hexagonal	4.10	Organic hybridic micelles dried at 60° C
L-α-PGA/C_{16}TAB/TEOS/ uncalc	Hexagonal	4.31	Inorganic-Organic mesophases dried at 60° C
L-α-PGA/C_{16}TAB/TEOS/ calcined	Hexagonal	4.13	Silicate mesostructures calcined at 650° C

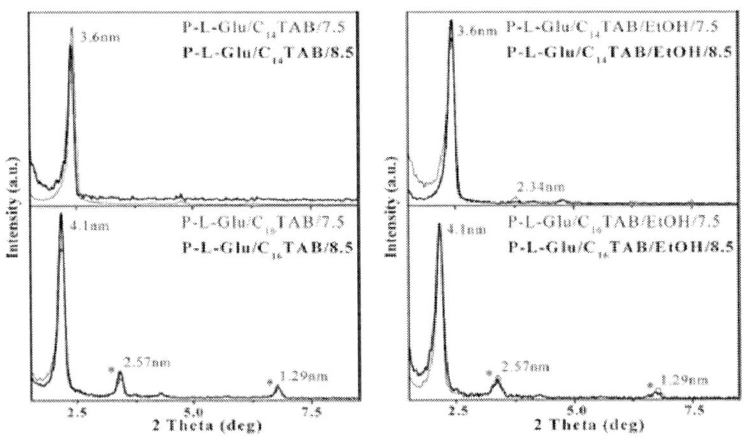

Figure 11. Left part: XRD patterns of the samples L-PGA/C_{14}TAB (upper part) and L-PGA/C_{16}TAB (lower part) isolated at pH values 7.5 and 8.5. Right part: Similar to those in the left part but with the addition of 0.5% EtOH. The peaks indicated by stars correspond to C_{16}TAB itself. Reprinted with permission from [2]. Copyright 2012 American Chemical Society.

The corresponding XRD –diffractograms are in Figure 11 for the samples without TEOS and in Figure 12 in the samples with TEOS. From the data in Figures 11 and 12, and some which will follow, the following observations can be made about the materials shown in Table 1:

(i) All the obtained materials (e.g. the Organic hybridic micelles PGA + C_nTAB dried at 60° C; the uncalcined Inorganic-Organic meso-phases PGA + C_nTAB +TEOS dried at 60° C and the calcined Silicate mesostructures PGA + C_nTAB +TEOS calcined at 650° C) show a blueprint corresponding to MCM-41 type structures –see inset in Figure 10.

Those structures are made up of cylindrical micelles, or cylindrical porous structures, packed in hexagonal fashion reminding spaggeti. A depiction of the mechanism leading to such structures is shown in the same Figure 10. It seems that the first systematic attempt to study the interaction of acidic poly-aminoacids, like Poly-Aspartic Acid (PAA) and PGA, with cationic surfactants C_nTAB (n=12 to 22) has been made by Munoz-Guerra and co-workers [27-29]. These authors have also proposed a model for the configuration of PAA-C_nTAB and PGA-C_nTAB complexes where the poly-aminoacidic backbones interact to each other and stabilized via the hydrophobic chains of surfactants species which form crystalline paraffinic phases. Obviously this model is similar to the one proposed for the interaction between PAC and C_nTAB [8, 9].

(ii) The distance between the centers of cylindrical micelles (dried materials) or the centers of silica mesopores (calcined solids) is 3.6nm for the PGA + $C_{14}TAB$ micelles and 4.1nm for the PGA + $C_{16}TAB$ micelles. If the above model (Figure 10) is correct, then this difference of 0.5nm (5Å) corresponds to *twice* the length of the additional group i.e. 2(–CH_2-CH_2 –) which means 0.12nm (1.2Å) per (– CH_2 –) group.

Then the distance of 3.6nm corresponds to 3.6/0.12= 30 (– CH_2 –) groups which tallies exactly to twice the length of the surfactant $C_{14}H_{29+1}N^+(CH_3)_3$ – e.g. (14C+1N) bonds. Similarly the distance of 4.1nm corresponds to 4.1/0.12= 34 (– CH_2 –) groups which corresponds to twice the length of the surfactant $C_{16}H_{31+1}N^+(CH_3)_3$ – e.g. (16C+1N) bonds.

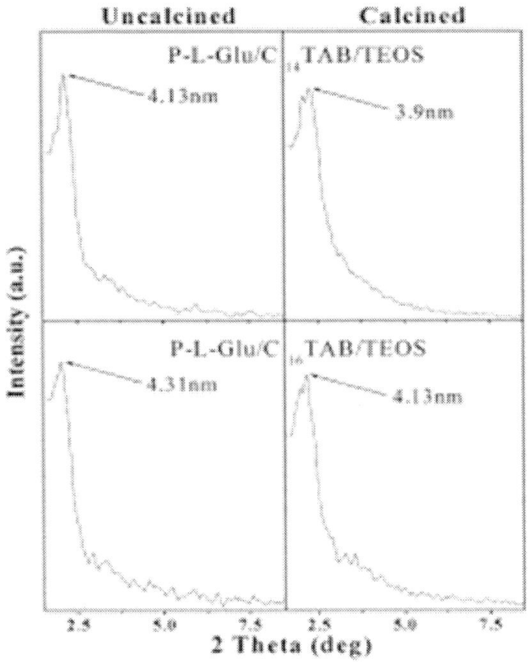

Figure 12. XRD patterns of the samples L-PGA/C_{14}TAB/TEOS (upper part) and L-PGA/C_{16}TAB/TEOS (lower part). Left: Uncalcined samples; Right: Calcined samples. Reprinted with permission from [12]. Copyright 2012 American Chemical Society.

(iii) The shrinking observed between the *micelles* L-α-PGA/C_{16}TAB/TEOS/uncalc (dried at 60°C) and the mesopores in the *silicate material* L-α-PGA/C_{16}TAB/TEOS/calc (fired at 650°C) equals 4.31-4.13=0.19 nm. For the *micelles* L-α-PGA/C_{14}TAB/TEOS/uncalc and the mesopores in the *silicate material* L-α-PGA/ C_{16}TAB/ TEOS/calc the shrinking is 4.13-3.90=0.23nm. This shrinking should be due mainly to the removal of organics from the samples fired at 650°C. The thermal behavior of the hybridic materials L-α-PGA/C_{16}TAB/TEOS/uncalc is shown in Figure 13 in the form of Thermo-Gravimetric (TG) and Differential Thermal Analysis (DTA) curves. The *endothermic* weight loss ~8% in the temperature range up to ~100° C should be due to the removal adsorbed of water. Then the *exothermic* weight loss in the temperature range 200-650° C should be due to the burning of organics. The remaining inorganic phase accounts for about 30% of the original dried mass.

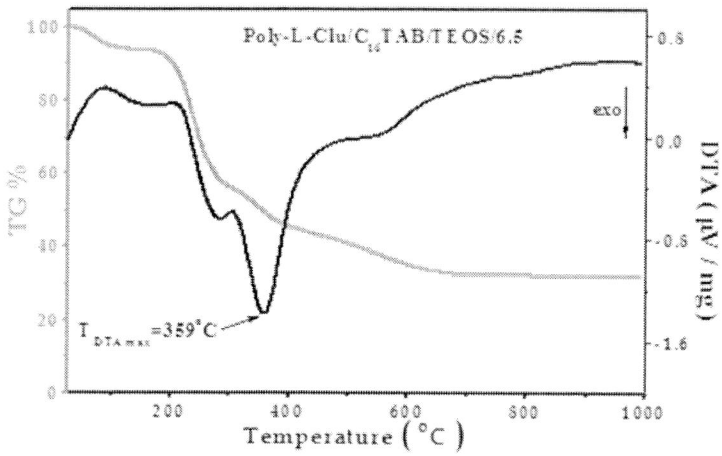

Figure 13. TG and DTA of the sample L-PGA/C_{16}TAB/TEOS isolated at pH=6.5.

Assuming in a schematic way that the hybridic material is built up of units of the form

X[Si-O] - *Y*[-CO-NH-CH (CH_2 -CH_2 -OCO-ATC_{16})-NH-]

it can be easily verified that X:Y should obtain values around 4.0-4.5. In other words about four to five layers of [Si-O] units cover each micelle. The formation of such micellar aggregates could be imagined along the following reaction path.

$$4(\mathrm{Si}\text{-}[\mathrm{OEt}]_4) + 16\mathrm{H_2O} + [-\mathrm{CO-NH-CH-CH_2-CH_2-OCO-ATC_{16}}] \rightarrow 4\mathrm{Si(OH)_4}-[-\mathrm{CO-NH-CH-CH_2-CH_2-OCO-ATC_{16}}] + 16\mathrm{EtOH}$$

It should be kept in mind that the above reaction is schematic and does not include the condensation of the hydro-silicate species in the sites on which they are attached on the soft organic backbone, a problem unsolved for the moment. Nevertheless, if two hybridic micellar aggregates are in contact to each other, the combined silicate wall/shell should be about 8-10 [Si-O] units thick, corresponding to 12-15 nm as observed experimentally in the usual MCM materials.

(iv) In Figure 14 the N_2 adsorption-desorption isotherms (77K) for the silicate porous materials L-α-PGA/C_{14}TAB/TEOS/calcin and L-α-PGA/C_{16}TAB/TEOS/calcin are shown together with the pore size distribution (psd) estimated according to the Horvath-Kawazoe (HK) method. The estimated specific surface area S_p (BET) (m^2/g), the specific pore volume V_p(cm^3/g) and the maximum of the pore size distridution d_{max} (nm) is shown in Table 2.

Table 2. Specific surface area S_p (BET) (m^2/g), Specific pore volume V_p(cm^3/g), Maximum of the pore size distridution d_{max} (nm) of sample PGA/C_{14}TAB/TEOS/calcin and PGA/C_{16}TAB/TEOS/calcin. The porosity is mixed, ordered and random

Sample	$S_{p,}$ (BET) (m^2/g)	V_p (cm^3/g)	Max of psd,d_{max} (nm)
L-α-PGA/C_{14}TAB/TEOS/calcin	1076	0.89	2.46
L-α-PGA/C_{16}TAB/TEOS/calcin	1001	1.12	2.60

From the shape of isotherms in Figure 14, especially the knee at (P/P_0) ≈ 0.35 it is assumed that there is some ordered porosity of MCM-41 type as indeed shown and by XRD data in Figure 11. But clearly the MCM-41 type order is much deteriorated and there is also extented non- ordered, or random(worm-like) porosity. The surface area of the obtained solids is in the range of ≥1000(m^2/g) which is usual for this kind of solids.

(v) The deterioration of the organization in the calcined samples L-PGA/CTAB/TEOS compared to the corresponding uncalcined samples (compare Figure 11 and Figure 12) can be clearly seen in SEM photography: In Figure 15 there are some typical SEM photographs of the calcined samples of P-L-Glu/C_{14}TAB/TEOS. In Figure 16 there are similar photographs of the uncalcined samples P-L-Glu/C_{16}TAB/TEOS and P-L-Glu/C_{14}TAB/TEOS.

It is observed that the uncalcined samples (Figure 16) show a clear directional development and structure which is lost is the calcined materials (Figure 15) that are essentially amorphous silica SiO_x. This transformation is due to the thermal removal of ~70% of the organic scaffold of the system (see

Figure 13) which leaves back the macroscopically amorphous silicate structure. Nevertheless this silicate material bears to some extent the inprinting features of its ordered hydridic precursor, e.g. partially ordered mesoporosity, as seen in the N_2 adsorption- desorption isotherms in Figure 14.

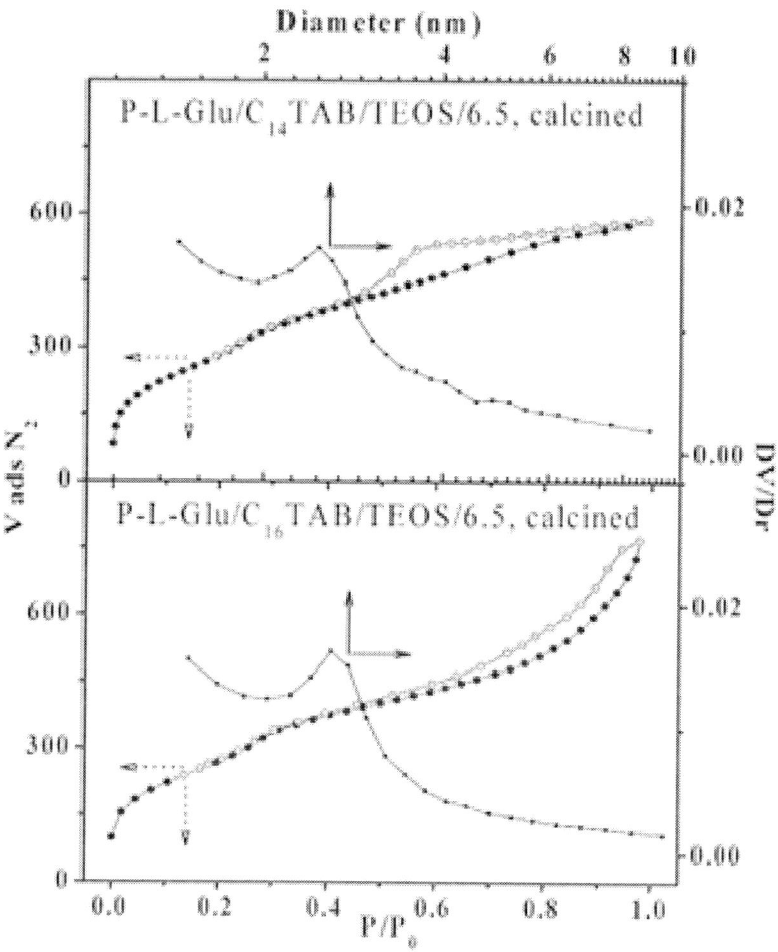

Figure 14. Nitrogen adsorption-desorption isotherms V=f (P/P_0) (left and low axes) and thr corresponding pore size distribution DV/Dr =f(d) (right and upper axes) for the samples L-α-PGA/C_{14}TAB/TEOS/calcin (upper part) and L-α-PGA/C_{16}TAB/TEOS/calcin (lower part).Reprinted with permission from [2]. Copyright 2012 American Chemical Society.

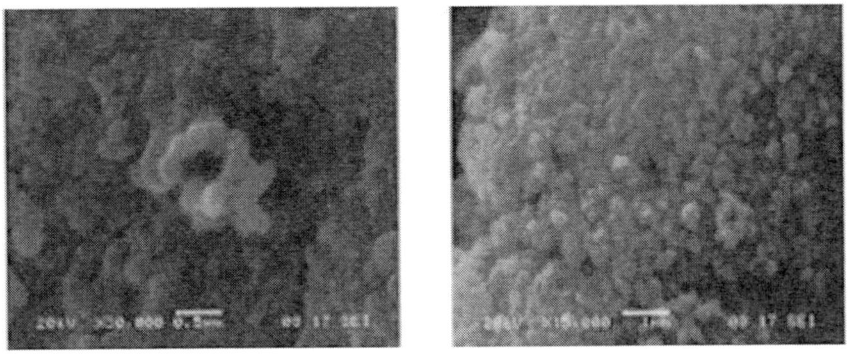

Figure 15. Typical SEM photos of P-L-Glu/C_{16}TAB/TEOS/calcined (left) and P-L-Glu/C_{14}TAB/TEOS/calcined (right).

A careful observation of the directional texture of the uncalcined samples, observed in the successive SEM microphotographs in Figure 16, reveals that it is somehow developed at three successive scales of weaving:

- The first scale (A) can be seen in the two upper photos in Figure 16. It shows parallel wrinkles extended for several tens, or even hundreds, of micro-meters. Let name them "1st order wrinkles" and the axis of their main direction let be (X||).
- The second scale (B) is shown in the two middle photos in Figure 16. It can be seen that the extended 1st order wrinkles bear directional (non-random) anomalies, which correspond to "2nd order wrinkles", and are positioned *vertically to the X-axis* (X⊥).
- The third scale (C) is shown in the two lower photos in Figure 16. It can be seen that the 2nd order wrinkles bear some kind of parallel 3rd order fine weaving vertical to the (X⊥) axis and therefore almost in the same direction of (X||) axis. This last fine weaving can be clearly seen in Figure 17, which is the enlarged version of the low-left Figure of Figure 16.

How these successive scales of organization, normal to each other, can be approach using some kind of prototype or model? Are there similar natural or artificial structures developed in a similar fashion? A positive answer to this may be related to the cross-linked model proposed for the γ-polyglutamic acid which is shown in Figure 18. The formation of γ-PGA starts with a γ-carboxyl group linked with α-amino group (see Figure 6). Then the γ-PGA chains are linked to each other towards a very high molecular weight polymeric mesh.

The obtained mesh structure has a molecular weight of tens of millions according to the commercial supplier of this product Nippon Poly-Glu Co., Ltd. A model similar to the cross-linked mesh, shown in Figure 18 for the γ-PGA, can be probably used to approach the fine parallel weaving seen in Figure 17 for the α-PGA/CTAB/TEOS.

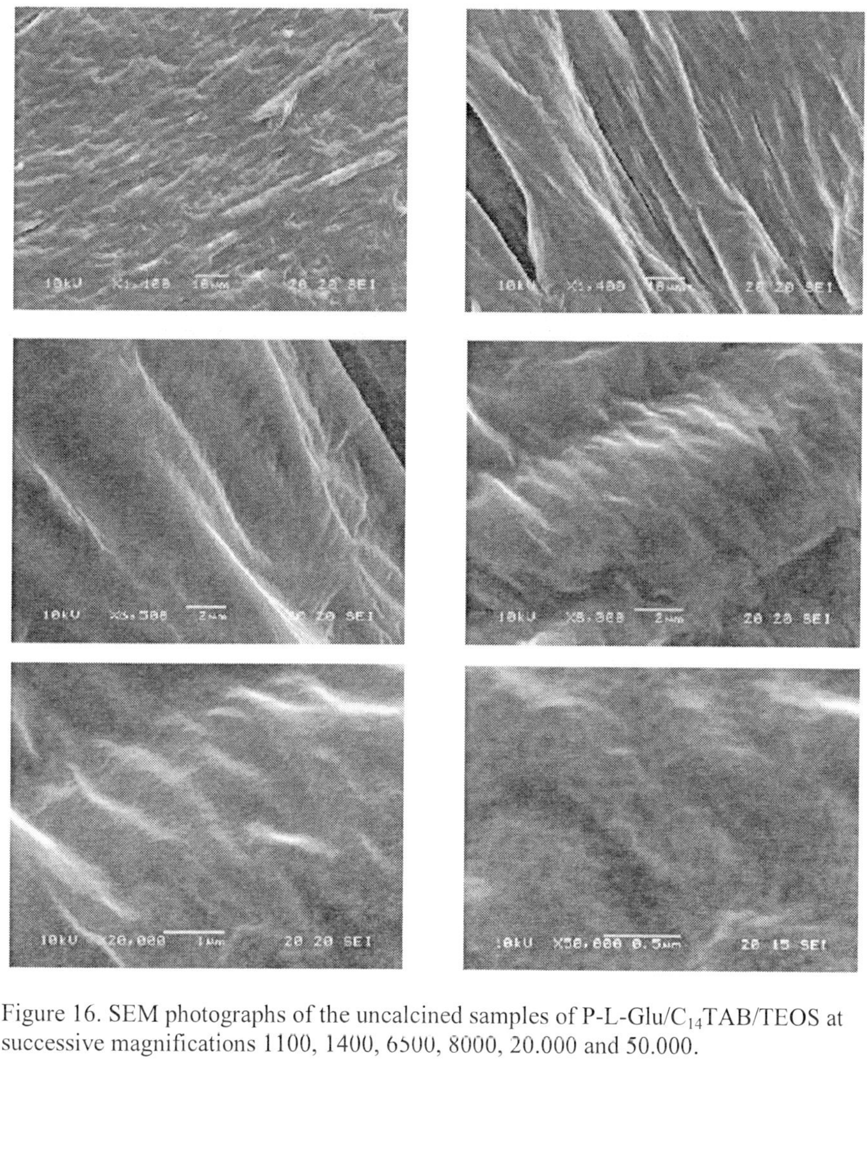

Figure 16. SEM photographs of the uncalcined samples of P-L-Glu/C_{14}TAB/TEOS at successive magnifications 1100, 1400, 6500, 8000, 20.000 and 50.000.

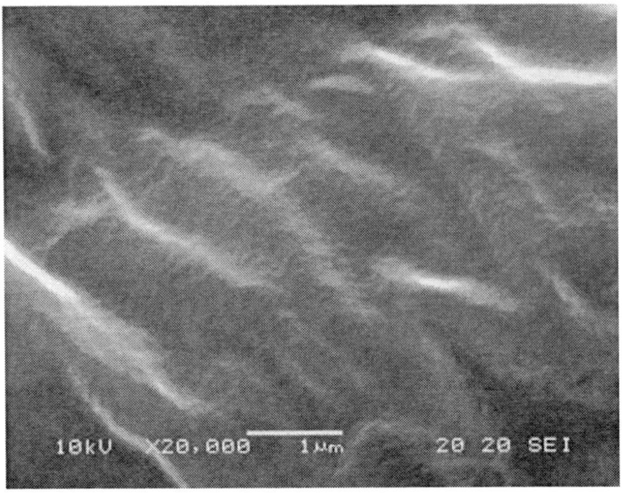

Figure 17. SEM photographs showing the fine weaving of the uncalcined samples of P-L-Glu/C_{14}TAB/TEOS (magnification 20.000). This Figure is an enlarged version of the low-left Figure of Figure 16. The parallel treads of the fine weaving are better seen in the lower right hand corner.

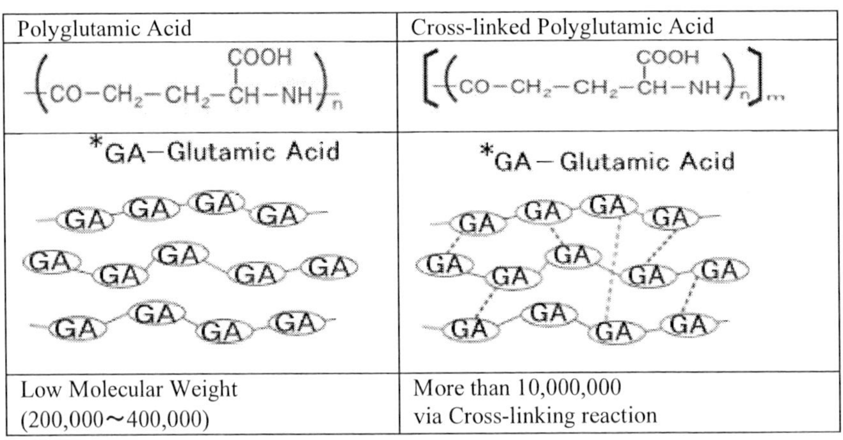

Figure 18. Cross-linked γ-polyglutamic acid consists of mesh structure of tens of millions by molecular weight. The formation of γ-PGA starts with a γ-carboxyl group linked with α-amino group (see Figure 6). Then the γ-PGA chains are linked to each other towards a very high molecular weight polymeric mesh. Compared tonon-linked PGA, the cross-linked PGA has higher water absorption capability (from Nippon Poly-Glu Co., Ltd).

4. Structure of Enantiomeric Hybridic Phases (L-PGA + CTAB) and (D-PGA + CTAB)

The purpose of this section is to search for the structure of the enantiomeric hybrid phases L-α-Poly-Glutamic Acid+CTAB) on one hand and (D-α-Poly-Glutamic Acid +CTAB) on the other. As a consequence we shall also attempt a modeling of the obtained structures-see previous section. The searching of the enantiomeric structures leads to some very fundamental questions concerning the enigmatic presence of L-isomers in nature. Let us be more specific.

The helical forms of proteins. For reasons which have not been resolved yet, the aminoacids used as building blocks of all proteins in various organisms are the left-hand isomers or the L-type ones [30–41]. The resulting macromolecular proteins, which are the cornerstones of life, obtain usually - but not always - the conformation of *right-hand or α-helix* [42–49]. So, the development of right-hand α-helix is the result of successive bonding of L-type aminoacids to each other via peptide –CO–HN– bonds (see Figure 2). There are 3.6 aminoacid residues per turn of the α-helix. Each turn covers 0.54 nm along the axis of the helix and each residue 0.15 nm. The helicity is generated and stabilized by hydrogen bonds between ==CO and HN== groups of i and the i+4 aminoacids in the chain. The weak hydrogen bond (≈15–25 kJ/mol) results in a formation of flexible 13-member-rings [49]. The 13th member of the ring is the hydrogen of the homonymous bond itself. This is the reason sometimes this helix is termed 3.6_{13}-helix, meaning that a 13member-ring is formed every 3.6 aminoacid residues. Apart from this common α-helix, various polyaminoacids can also adopt the form of the 3_{10} *-helix*. This is a similar right-hand helix, but now the hydrogen bond results in the formation of a 10-member-ring [49]. There are exactly three residues per turn of the helix and the 10th member of the ring is the hydrogen atom itself. The hydrogen bonds between ==CO and HN== groups are now formed between the i and the i +3 aminoacids of the chain. The 3_{10}-helix is less tightly packed and more extented and unfolded compared to α-helix. Molecular dynamics calculations in water have shown that the α-helix unfolding may take place via a 3_{10} - intermediate [50]. The α-helix possesses an inherent higher structural stability, compared to its 3_{10}-counterpart. A substantial amount of all 3_{10} helices occur at the ends of alpha-helices. There are also two very rare additional helical structures in proteins, the more packed pi-helix (which is an extremely rare secondary structure) and the less packed 2_7 ribbon but those structures are not of interest in the following discussion. Details about them are included in Table 3 and shown in Figure 19. There are three common secondary structures in proteins, namely *alpha helices, beta sheets and turns.* That which cannot be classified as one of the standard three classes is usually grouped into a category called *"random coil".* This designation is unfortunate as no portion of protein three dimensional structure is truly random and it is not a coil either. Alpha-helix, beta-sheet and random coil structures each give rise to a characteristic shape and magnitude of the *Circular Dichroism (CD) spectrum* as shown in Figure 20.

Table 3. Parameters of regular secondary helical structures. n is the number of residues per helical turn, p is the helical pitch and A is the atoms in H-bonded loop

Structure	n	p(Å)	A	H-bond (CO, HN)
3_{10}-helix	3.0	6.0	10	i, i+3
alpha helix [3.6_{13} - helix]	3.6	5.4	13	i, i+4
pi-helix [4.4_{16}- helix]	4.4	5.0	16	i, i+5

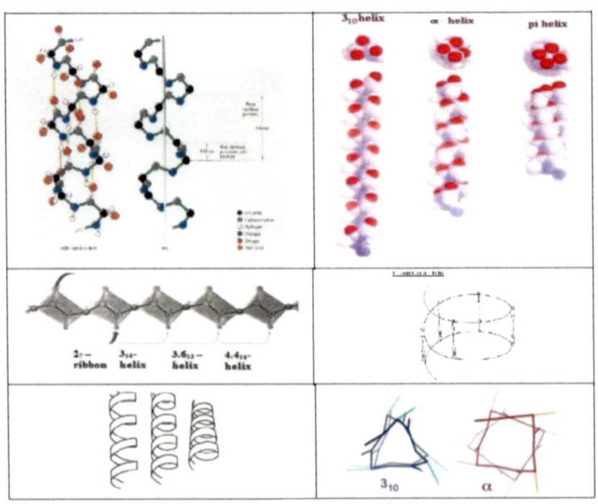

Figure 19. Upper left: The alpha helix in a typical polypeptide sequence. Upper right: The 3_{10}-helix, the alpha helix (or 3.6_{13} – helix) and the pi-helix (4.4_{16}-helix). Middle left: Schematic drawing of hydrogen bonding which stabilizes the 2_7–ribbon, the 3_{10}-helix, the alpha helix (3.6_{13}–helix) and the pi-helix (4.4_{16}-helix). Middle right: The H-bonds along the helix. Lower left: Representation of the left hand helices suffering gradual transformation from a loose to a more tight configureuration. Lower right: Contrast of helix end views between α- (offset squarish) vs 3_{10} (triangular) helices.

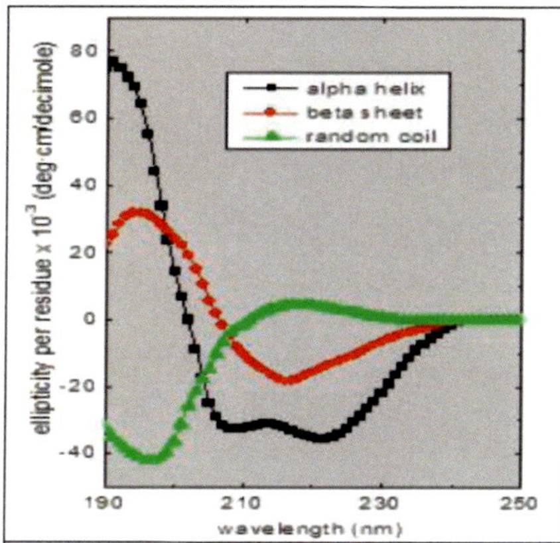

Kind of configuratiom	Max	Min
Alpha helix	~190	~208 ~222
Random coil	~220	~200
Beta sheet	~198	~215

Figure 20. The standard curves used to determine the secondary structure of proteins using CD spectroscopy in the "far-UV" spectral region (190-250 nm). At these wavelengths the chromophore is the peptide bond, and the signal arises when it is located in the indicated regular environment. Alpha-helix, beta-sheet, and random coil structures each give rise to a characteristic shape and magnitude of CD spectrum. The approximate fraction of each secondary structure type that is present in any protein can thus be determined by analyzing its far-UV CD spectrum as a sum of fractional multiples of such reference spectra for each structural type. Like all spectroscopic techniques, the CD signal reflects an average of the entire molecular population.

The alpha-helix and beta-sheets conformations for polypeptide chains are the most thermodynamically stable of the regular secondary structures. However, particular amino acid sequences of a primary structure in a protein may support regular conformations of the polypeptide chain other than alpha-helical or beta-structure. Thus, whereas alpha-helical or beta-structure are found most commonly, the actual conformation is dependent on the particular physical properties generated by the sequence present in the polypeptide chain and the solution conditions in which the protein is dissolved.

The 3_{10} helix contains one additional H-bond compared to α-helix for the same peptide length and is better entropically stabilized. In solution the inherent increased stability of the α-helix often dominates the energy contribution of the single additional hydrogen bond formed in a 3_{10}-helix [51], but the balance is delicate and can be altered, for example, by side-chain effects [51d] or the polarity of the solvent [51e].

The carboxyl (or C-terminal) and the amino (or N-terminal) ends of the polyamino-α-chains obtain often the 3_{10} –helical form, in other words they suffer a subtle, not destructive transformation due to the solvation effects of water [52, 53]. The existence of such 3_{10} -helical ends in a polypeptide chain has some important biochemical consequences related to the antimicrobial action of polypeptides [54, 55].

The Circular Dichroism of helical forms. The two main kinds of helices, the α–helix and the 3_{10}–helix, bear very distinct fingertips in the *Circular Dichroism (CD) spectra* [44, 46, 56–59]. Namely, they adsorb light in the UV range as follows:

- The α–helix adsorbs light at 222–225 nm via a (n →π*) transition.
- The 3_{10}–helix adsorbs light at 205–209 nm via a (π →π*) transition).

The relative intensity of adsorption of the polarized light is expressed as molar ellipticity [θ] in units (deg cm ^{2} / dmol) – see Figure 20. If the ratio R = [θ] (n → π*) / [θ] (π → π*) tends to unity or larger values, the helix is of the α-form. If, on the other hand, the ratio R tends to 0.4, the helix is of the 3_{10}-form. Clearly, if this ratio obtains intermediate values, the total helical content is a mixture of α-helix and the 3_{10}-form.

Such chiral molecules, i.e., right-hand α- and 3_{10}–helices made up of L-aminoacids and their counterparts, i.e., left-hand and 3_{10}-helices made up of D-aminoacids, can be used principle as stereoselective homogeneous adsorbents, sensors and catalysts [60–62]. The complexation of such enantiomers and the formation of micellar mesophases using various cationic and/or anionic surfactants, like C_nTAB [63] and/or SDS [64], provides a convenient method to stabilize such materials.

So the purpose of the present section is to study the complexes of L-PGA/CTAB and D-PGA/CTAB and establishing their chiral conformation CD spectra. The second aim is examining the structure of the resulting micellar aggregates by X-ray diffraction analysis for the presence, or lack, of order. A third very important target is the examination of the thermal and hydrolytic stability of the chiral complexes L-PGA/CTAB and D-PGA/CTAB, as well as of the original free L-PGA-and D-PGA in the temperature range 10–70 °C using their CD spectra as a probing method. This method provides the opportunity to distinguish between the α- and the 3_{10} -helices and their relative development as a function of temperature. The results about the stability of helices will be discussed in the next section in relation to the possible selective action of the spin isomers of water on them.

Sample pereparation and methodologies. The synthesis of the hybridic materials is described in details in refs [2, 3]. The polyaminoacids used were L-α- PGA (MW = 5000–15000 from Sigma–Aldrich) and D-α-PGA (MW = 5000–15000 from Sigma–Aldrich). The complexation with the surfactant

CTAB (see Figure 8) was such that there was a correspondence 1:1 between the molecules of the surfactant and the carboxyl groups of the polyaminoacids, using their mean MW = 10000. The obtained solution was acidified by the addition of 0.1 N HCl to pH 1.5. Then a slow titration of that acid solution started with 0.1 N NH_3 using an automatic Radiometer Copenhagen system. It was possible to stop the titration at selected pH values in order to isolate samples for characterization. The samples obtained at pH 8.5 were mostly used since at this pH the maximum complexation of the polyaminoacid with the surfactant takes place (see Figure 3). The samples thus isolated were then dried at room temperature and characterized by X-ray diffraction (XRD) analysis for their structure (see Figure 21) and circular dichroism (CD) for the conformation of the polypeptide–CTAB complexes (see Figure 22 and 23). The solutes used in the CD experiments were trifluoroethanol (TFE), either in pure form or in mixtures (v/v) with an aquatic solutions (termed E) of NH_3 and HCl of pH 8.5.

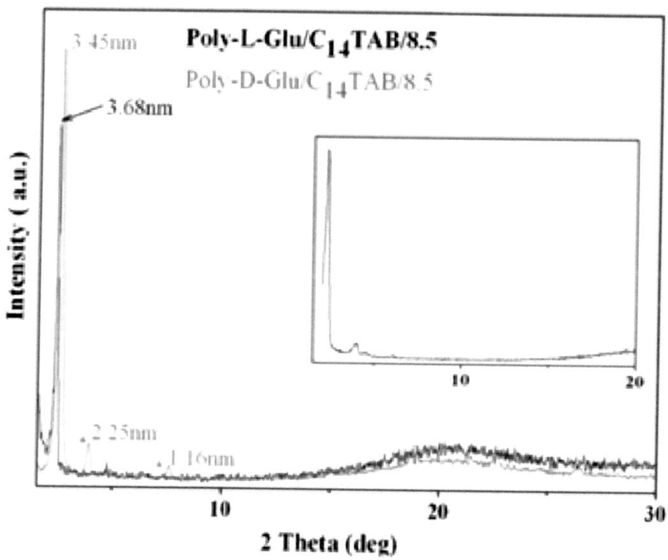

Figure 21. X-ray diffraction patterns of the complexed mesophases P-L-Glu/C_{14}TAB/8.5 and P-D-Glu/C_{14}TAB/8.5 isolated at pH 8.5. The small peaks indicated by stars correspond to free C_{14}TAB. The insct shows the XRD pattern corresponding to a typical MCM-41 material [10,13]. Reprinted with permission from [3]. Copyright 2012 Elsevier Science Publication.

The reasons for using TFE are discussed in ref [3]. Solution E was prepared by mixing H_2O, NH_3 and HCl to achieve pH 8.5 in a procedure exactly similar to that used for the sample preparation (see above) but without any polyaminoacid or CTAB.

The striking feature of the X-ray diffractograms of the complexed mesophases P-L-Glu/C_{14}TAB/8.5 and P-D-Glu/C_{14}TAB/8.5, shown in Figure 21, are the identical strong peaks around $2\theta \approx 1.2$ deg corresponding to Bragg diffracting levels at distance $d = 3.55 \pm 0.10$ nm. These patterns are similar to those shown in Figure 11 for the samples L-PGA/C_{14}TAB and L-PGA/C_{16}TAB as expected. This means that the X-rays probed micelles of similar ordered structures, made up of cylinders packed in a spaghetti-like fashion like the well-established MCM-41 materials.

The impressive point is that these are *chiral micelles* as shown clearly in the CD spectra in Figure 22 and 23: The mirror nature of the spectra indicates that one structure is a *left hand -α- helical micelle* and the other is its image, i.e. the *right hand-α- helical micelle.* Those micelles are depicted in Figure 24.

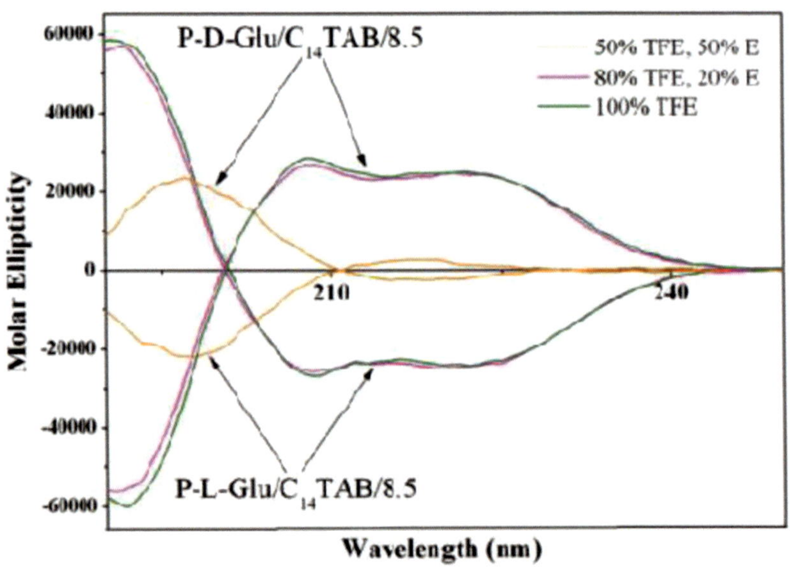

Figure 22. Circular dichroism (CD) spectra of the complexes P-L-Glu/C_{14}TAB/8. 5 and P-D-Glu/C_{14}TAB/8.5 in three different kinds of solvents made up of trifluoroethanol (TFE) and aquatic solution of HCl and NH_3 of pH 8.5 (E). Reprinted with permission from [3]. Copyright 2012 Elsevier Science Publication.

It is mentioned that those enantiomeric micelles exhibit *almost identical hydrolytic and thermal stability:* In Figure 22 it is observed that in mixtures TFE+H_2O the *left and/or right hand -α-helical micellar structures* are retained up to ratio 80 % TFE + 20 % H_2O. Then as the water content increases at 50 % TFE + 50 % H_2O the structures suffer some kind of hydrolytic destruction/destabilization and a transition to random coil (see standard curves in Figure 20).

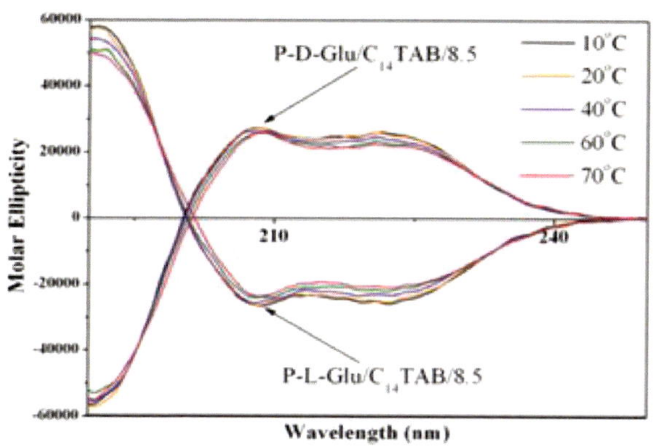

Figure 23. The CD spectra of the same complexes P-L-Glu/C_{14}TAB/8.5 and P-D-Glu/C_{14}TAB/8.5 at five different temperatures 10, 20, 40, 60 and 70 C in solvent 80% TFE–20% E. Reprinted with permission from [3]. Copyright 2012 Elsevier Science Publication.

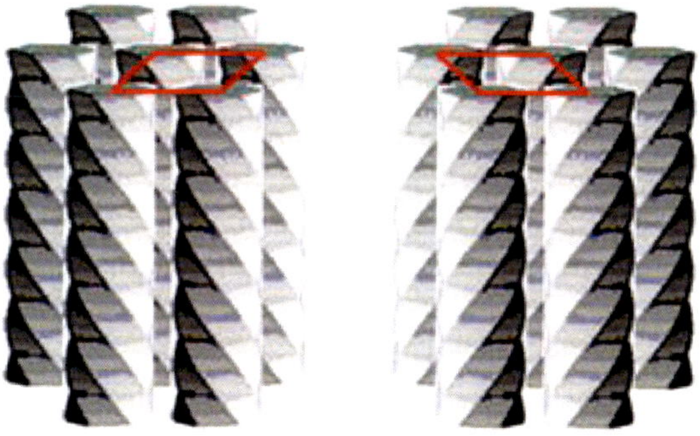

Figure 24. Depiction of the chiral micellar aggregates developed in the systems P-L-Glu/C_{14}TAB and P- D-Glu/C_{14}TAB packed in hexagonal lattices of similar size 3.55 ±0.10 nm (a,b-axes). Reprinted with permission from [3]. Copyright 2012 Elsevier Science Publication.

Similarly, as seen in Figure 22, both the *left and/or the right hand -α-helical micellar structures are thermally stable* in 80% TFE–20% E at least up to 70°C. It is possible now, using the above information, to propose a model at molecular level for the development of micelles. The model is biomimetic in the sense that it takes into account two well known examples from nature, namely the structure *of α-keratin* and of *collagen*, which are shown in Figure 25. The proposed model for the PGA/CTAB system is shown in Figure 26.

Both structures of α-keratin and of collagen start with a polypeptide α-chains. Then successive organization levels lead to the final hierarchically built structures- see Figure 25.

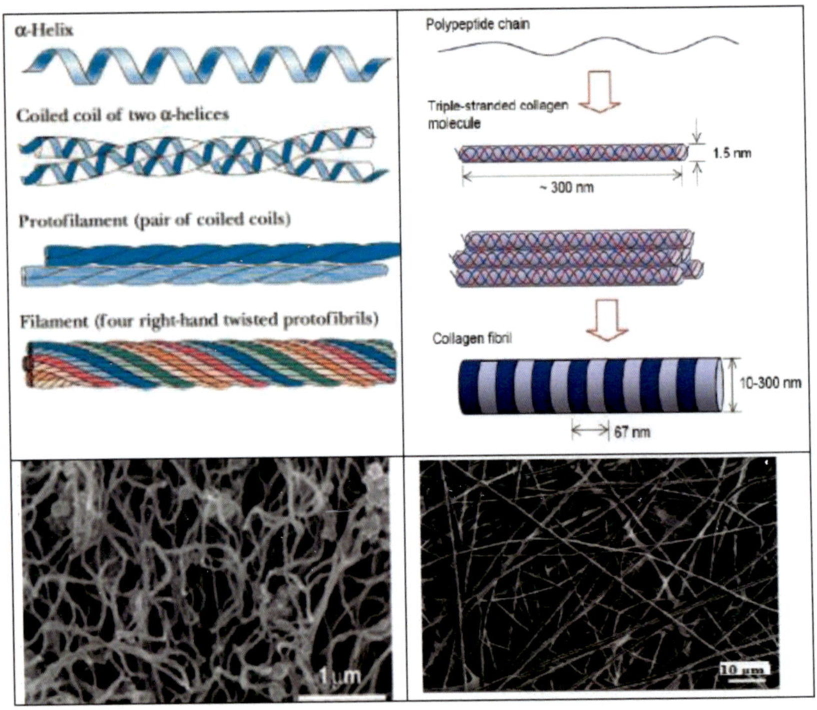

Figure 25. Left: The development of α–keratin (upper part) and its image in SEM (lower part). Right: The structure of collagen fibrils (upper part) and its image (lower part).

The *structure α-keratin* (key component of hair, wool, horns, nails, claws and hooves) starts with a single α-helix (Figure 25, left). Then two such coiled helices are coiled together. Next these coiled coils are approached to each other and form the ptotofilament or protofiblil.

Finally four right-hand protofibrils are twisted to each other forming cylindrical bundles structures. Such cylindrical structures are seen in SEM microphotography to have length of several micro-meters and thickness of 0.1-0.05 μm (100-50nm).

In a somehow similar scenario, the *development of collagen* (the main component of connective tissue and the most abundant protein in mammals) starts with a single peptide chain. Three such α-chains are twisted to a triple helix fibril. Those fibrils are then aligned laterally to form bundles of a higher order of structure and make up the tough micron-sized collagen fibres (see SEM image in Figure 25). A characteristic feature of the collagen fibrils is their banded structure. The diameter of the fibril changes slightly along the length, and as a result a highly reproducible band structure is repeated approximately every 67 nm. The thickness of fibrils is in the range 10-300nm.

The model for the structures PGA/CTAB of the present case should be somehow similar (Figure 26). Initially *cylindrical micelles, orderly packed in hexagonal fashion*, are obtained. The sefl-assembly is driven by the complexation of side acid groups of polyelectrolyte PAG with the cationic surfactant CTAB forming the backbone of the cylindrical micelle (see reaction 1). The structure is similar to MCM-41.

The hydrophilic head $C_nH_{2n+1}\,N^+(CH_3)_3$ of CTA must be located *outwards and towards the aquatic phase* where anionic oxy-hydroxy-silicate species from the hydrolysis of TEOS [e.g. $Si(OEt)_4 + H_2O \rightarrow Si(OH)_4 + 4EtOH$, $xSi(OH)_4 \rightarrow Si_x(OH)_yO_z^{-n} + kH_2O$) are formed and attracted/precipitated on the outer cylindrical surface (reaction 2).

The hexagonal packing is certified by XRD diffractograms and the Bragg refraction (100) at low angles ($2 < 2\theta < 3$) as shown in Figures 11 and 21. The calculated d-space corresponds to the distance between centers of the cylindrical micelles and is 3.5-3.6 nm for C_{14}TAB and ~4.1nm when C_{16}TAB is used.

The L-α-PGA or the D-α-PGA chains are connected to the hydrophilic head $-N^+(CH_3)_3$ of CTA which is located outwards and towards the aquatic phase (reaction 1). These chains tend to twist and wrapped around the micelle in the left-hand, or right –hand, direction respectively. So the micelles obtain enantiomeric structure. This becomes clear in the CD spectra shown in Figures 22 and 23. The enantiomeric micelles are shown in Figure 24. This behavior is in contrast to the usual MCM-41 materials which do not exhibit such optical activity. Such chiral micelles can be used principle as stereoselective adsorbents, sensors and catalysts [60-62].

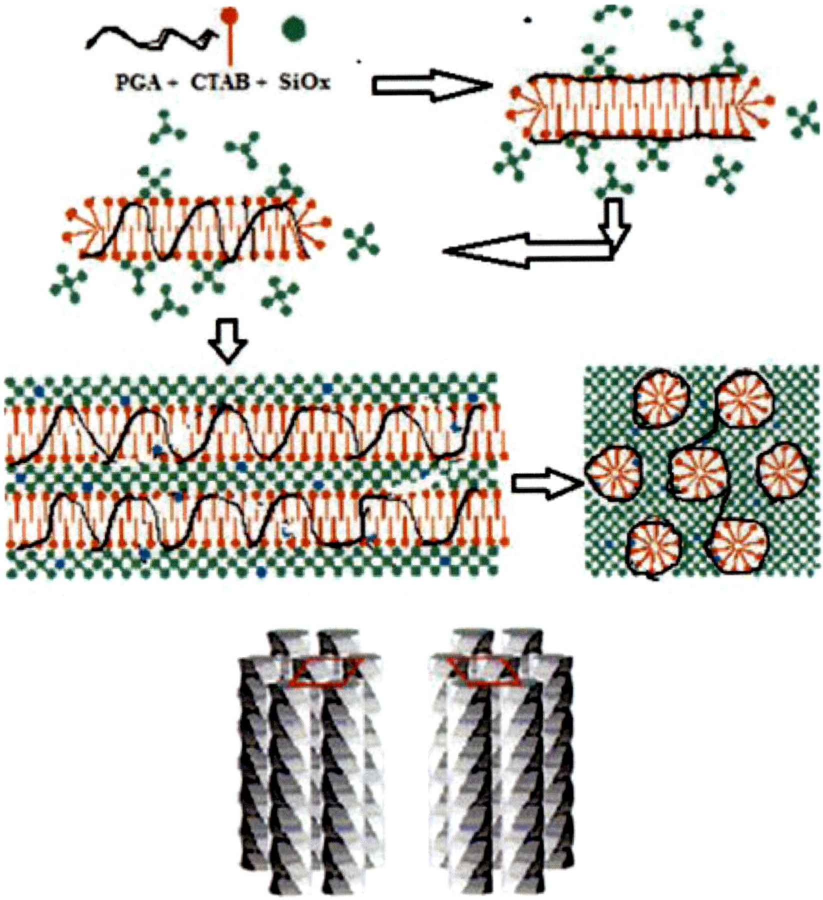

Figure 26. A possible mechanism for the development of the enantiomeric micellar mesophases L- PGA/CTAB /TEOS and D- PGA/CTAB /TEOS. The PGA wraps in a helical way the outside hydrophilic cylindrical micelles, hexagonally packed.

The single micelles, wrapped in left hand or right fashion by the L-PGA or D-PGA respectively, should have the natural tendency to form ordered bundles. It is suggested that such bundles are seen as fine weaving in the SEM picture shown in Figure 17.

A careful observation of this Figure reveals that those fine weavings have length of several μm and thickness of 30-50nm. This is the same order of magnitude for the fibers of α-keratin and of collagen shown in Figure 25. Since the primary micelles are 3.5-4.0 nm thick, the fine-weaving bundles should be made up of around 10-12 primary twisted cylindrical micelles.

5. Stability of Enantiomeric Structures (L-PGA Vs. D-PGA) and (L-PGA + CTAB Vs. D-PGA + CTAB)

The purpose of this section is to search for the relative *thermal stability* of the enantiomeric polyaminoacids (L-PGA vs. D-PGA), as well as of the enantiomeric hybridic phases (L-PGA/ CTAB vs. D-PGA+/CTAB). The study took place in an aquatic environment rich in trifluoroethanol (TFE) which protects the intrahelical H-bonds from hydrolytic attacks by water molecules. The experimental method used was the CD (see previous section and Figure 19) The CD spectra of the complexes P-L-Glu/C_{14}TAB/8.5 and P-D-Glu/C_{14}TAB/8.5 (Figure 22), were detected in three different kinds of solvents (100% TFE; 80% TFE and 20% E; 50–50% of TFE and E) and shows chiral helical forms induced by the presence of TFE, but destroyed by the addition of H_2O. Trivial calculations show [3] that when there are at least as many molecules of TFE as those of water, the helical structure is protected. When there is just one TFE molecule per four (4) water molecules, the proton-phobic shield protecting the helical form is destroyed. The ability of TFE to induce the formation of helical forms is well known and is due to the fact that its presence, in contrast to that of water, minimizes the formation of external hydrogen bonds of the polyaminoacid with water molecules existing in the hydration surface of the chain. So all the H-bonds are developed internally between either the i and the (i + 4) or the i and the (i + 3) aminoacid residues towards either α- or 3_{10} helices correspondingly. The above results indicate that, at least in the present case, a majority of water molecules (attacking the intrahelical H-bonds) is needed compared to TFE molecules (protecting the intrahelical H-bonds) for the transformation of the α- and/or 3_{10} -helices to random coil structure.

The CD spectra of the same complexes P-L-Glu/C_{14}TAB/8.5 and P-D-Glu/C_{14}TAB/8.5 detected at temperatures 10, 20, 40, 60 and 70°C in the solvent 80% TFE and 20% E (Figure 23) where the α- and/or 3_{10} -helices are preserved, as well as the similar CD spectra for the poly-L-and poly-D-glutamic acids themselves (Figure 27), i.e., without any C_{14}TAB groups attached to them, in the same solvent preserving the α- and/or 3_{10}-helices, show two absorption bands, one at 222–225 nm and a second at 205–209 nm. The intensity of the n → π* transition/absorption band at 222–225 nm is related to the content of the α-helix while that of π → π* transition/absorption band at 205–209 nm is related to the 3_{10}-helical form [15, 17, 27–30].

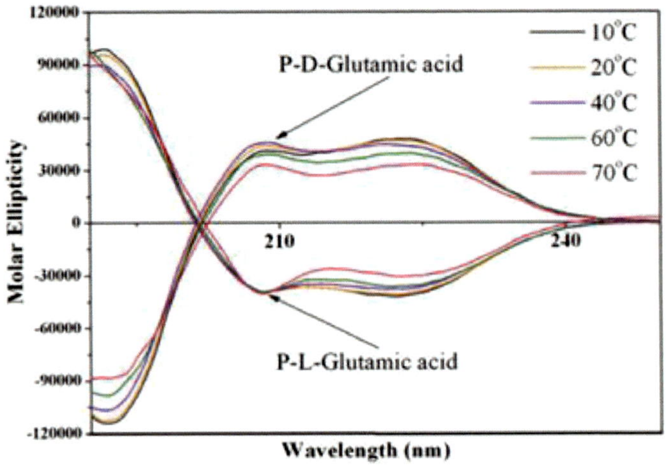

Figure 27. The CD spectra of P-L-and P-D-glutamic acids at five different temperatures 10, 20, 40, 60 and 70°C in solvent 80% TFE–20% E. The absorption band at 222 nm is due to a (n → π*) transition and is related to the content of the α-helix. The absorption band at 208 nm, is due to a (π → π*) transition and is related to the content of the 3_{10}-helix. Reprinted with permission from [3]. Copyright 2012 Elsevier Science Publication.

The intensity of absorption is expressed by the molar ellipticity [θ] in units (deg cm^2/dmol) and the ratio R = [θ] (n → π*)/[θ] (π → π*) can be used for the estimation of the two helical forms: For R ≥1, the helix is of the α-form, while for R ≈ 0.4 the helix has of the 3_{10}-configuration. Estimation of values of R have been carried out for the spectra in Figures 23 and 27, and the results are shown in Figure 28 in the form R = f(T). With one exception corresponding to the P-D-Glu/CTAB/40◦C sample, all the other data in this Figure 28 indicate that the ratio R = [θ] (n → π*)/[θ] (π → π*) ≈ (α-helix) / (3_{10}-helix) drops as the temperature increases from 10 to 70°C. This means that, either a more extensive part of α-helix is degenerated to random coil or that a more extensive part of α-helix suffers a gradual transition to 3_{10}-helical structure. As temperature increases, longer parts of the less tightly packed, more unfolded and entropically favored 3_{10} -helices are maybe formed, probably at the ends of the polyaminoacid chains, leading eventually to the total fraying of the helical structure according to the sequence: *α-helix→3_{10}-helix →random helix*. This transition α-helix → 3_{10}-helix appears more distinct and profound in the pair of the free polyaminoacids poly-L- and poly-D-glutamic acid, compared to the pair of the complexed chains P-L-Glu/C_{14}TAB and P-D-Glu/C_{14}TAB.

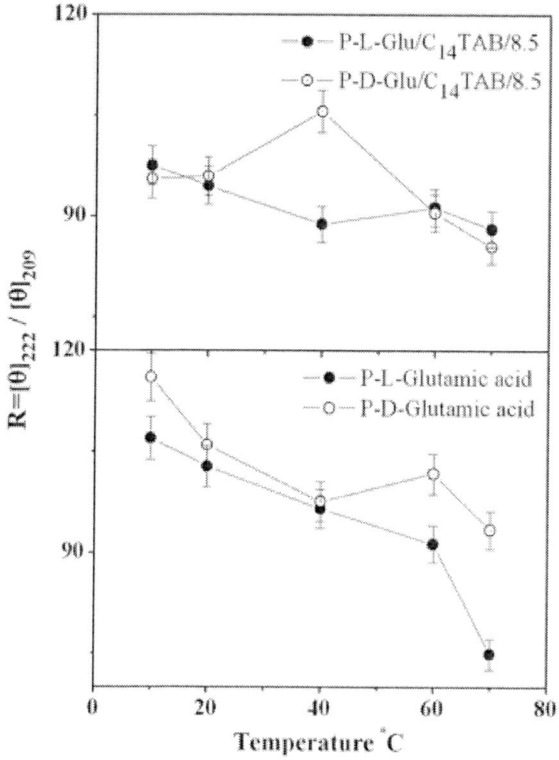

Figure 28. Variation of the ratio $R = [\theta]_{222} / [\theta]_{208}$ as a function of temperature for the indicated helical structures. The bars correspond to the experimental deviations. The lines are guide for the eye. Reprinted with permission from [3]. Copyright 2012 Elsevier Science Publication.

This should be due to the protective effect of surfactant. A second important point revealed from the comparison of the pair of the free polyaminoacids in Figure 28 is that the poly- D-glutamic acid, existing initially in the α-helix form, appears as more resistant towards thermal degeneration to random coil, or transformation to 3_{10}-helix, compared to the poly-L-glutamic acid. The same may be true for their corresponding protective forms, P-D-Glu/C_{14}TAB and P-L-Glu/C_{14}TAB, but in this case the verdict is not clear. To remove some of the ambiguities in the data shown in Figure 28, the reduced intensity $[\theta_{222}] / [\theta_{222max}]$ of the n $\rightarrow \pi^*$ absorption band, related to the content of the α-helix, as well as the reduced intensity $[\theta_{208}] / [\theta_{208\,max}]$ of the $\pi \rightarrow \pi^*$ absorption band, related to the content of the 3_{10}-helix, were estimated. Those ratios express how the two helices in each particular sample increase or decrease, not in relation to each other but in relation to their own initial conformation. The results are shown in Figure 29 as a function of temperature. A careful examination of the trends in Figure 29 leads to the conclusion that the content of α-helices in all structures drops faster compared to the content of the 3_{10}-helices. More important is the fact that in the poly-L-glutamic acid the content of the 3_{10}-helix remains constant, while that of the α-helix decreases fast. *The 3_{10}-helix of the L- structures appears as more resilient to temperature fluctuations.* In the case of poly-D-glutamic the content of both 3_{10} - and α-helices decreases with temperature in a, more or less, uniform way.

To summarize, the most important experimental findings from the CD spectra are:

- The D-isomers appear as more thermally stable, compared to L-isomers, both for the free polyaminoacids and their complexed forms with CTAB surfactants
- (ii) The 3_{10}-helix appears as more thermaly stable, compared to the α-helix, in the case of the L-isomer of the free polyaminoacid.

The discrimination of the thermal stabilities of the poly L- and the poly-D-glutamic acid is in line with some recent findings by Scolnik et al. [12], who studied the same systems by CD and isothermal titration calorimetry (ITC). They found differences in the stabilities of the two isomers, as a function of pH, for the transition α-helix → random coil when the solvent H_2O as substituted by D_2O.

The poly-D-glutamic acid showed a stronger tendency to keep its α-helix form and it was assumed a higher α-helix stability towards random coil degeneration in the pH range 8.2 to 3.5. In [12] no discrimination between the 3_{10} - and α-helices was attempted.

The explanation proposed was that the poly- L-glutamic in water interacts and is solvated slightly more than its mirror image poly-D-isomer. *This differentiation was attributed to the ortho-H_2O, constituting 75% of the usual water compared to 25% of the para-H_2O*, [48, 49]. This conjunction, and its far reaching consequences, will be discussed in the next section.

In any case, and independently of the reason behind the observed differences, the results of the present study support and extend the conjunction proposed by the group of Shinitzky [12] that the helical forms of the D-configurations appear more stable, not only in pH disturbances studied in [12], but also in temperature generated fluctuations.

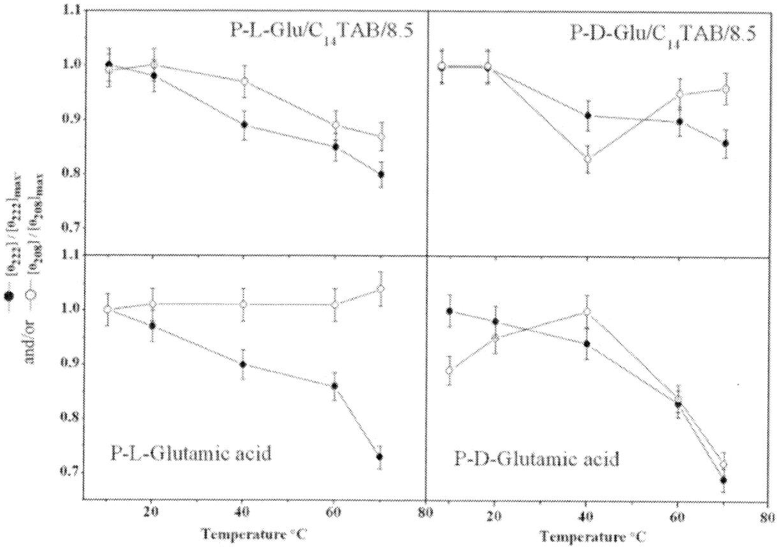

Figure 29. Variation of the reduced intensities $[\theta_{222}] / [\theta_{222\ max}]$ of the (n →π*) absorption band at 222 nm, related to the content of the α-helix, and $[\theta_{208}] / [\theta_{208max}]$ of the (π →π*) absorption band at 208 nm, related to the content of the 3_{10}-helix, as a function of temperature for the indicated helical structures. The bars correspond to the experimental deviations. The lines are guide for the eye. Reprinted with permission from [3]. Copyright 2012 Elsevier Science Publication.

6. The Subtle Role of Water

The purpose of this section is to discuss the possible role of water, and its spin isomers ortho-water and para-water, in the hydrolysis and the stability of D- and L- forms of poly-aminoacids. Such questions are eventually related to the enigmatic presence of L- isomers in the proteins and the total exclusion of D-isomers from them. The possible role of water is of paramount importance since the interaction of the polypeptides and proteins with their surroundings, the stabilization of α- or 3_{10} -helical structures, and their conversion from one to the other, take place practically in aquatic environment.

It is not probably well-known, but it is well established, that water exists in nature in two spin isomers, the ortho-H_2O (o-H_2O) and the para-H_2O, (p-H_2O) in a 3:1 ratio. They are discriminated by the alternative orientation of the hydrogen atoms nuclear spins-either parallel (total spin I = 1 and nuclear statistical weight 2I+1 = 3 for the paramagnetic o-H_2O molecule) or antiparallel (total spin I = 0 and nuclear statistical weight 2I+1 =1 for the nonmagnetic p-H_2O molecule). The aquatic environment in the vicinity of bio-polymers is actually deceptively simple: The hydrogen bonds themselves, their arrangement near the surface of them, the possible clustering of water molecules in the helical cavities and the possible effect of the o- and p-water on those bio-physicochemical processes, constitute a canvas of fine endless questions.

In Ref. [4, see also 41] Shinitzky et al. attributed the increased solvation of the poly-L-glutamic acid, compared to poly-D-glutamic acid, to the o-H_2O existing in a 3:1 excess in the usual water [48,49]. The preference of o-H_2O as a selective solvent of the L-isomers was attributed to the fact that it bears a weak magnetic field, which favors its interaction with the parity violation energy difference (PVED) of the L-enantiomers. This difference is originated from the electro-weak nuclear force, which is chiral [30, 67]. This chirality has been proposed to favor the selective interaction with the L-isomers. For example, Mason and Tranter [68] using a Hamiltonian with a weak interaction term, showed that the L-alanine is more stable than the D-alanine but the energy difference is only 10^{-19} eV/mol. It has been proposed [3, 4] that this, otherwise extremely weak effect, is amplified by autocatalytic helix formation in the sense that each turn catalyzes the formation of the next one. So, there is no need for the PVED to participate in the creation of the whole right-hand helix: The formation of just a single right-hand turn somewhere in the chain, generates the propagation of the whole helix. To use a mechanical metaphor, a slight externally forced turn at the end of a string, either clockwise or anticlockwise, forces the whole string to obtain the corresponding helical clockwise or anticlockwise structure. Furthermore, if the external turn is strong, the string becomes more twisted (α-helix?), while if the external turn is weak, the string becomes less twisted (3_{10} -helix?).

A few years ago Tikhonov and Volkov were able to show that water adsorbed on hydrophobic surfaces like carbon [65] and crosslinked polystyrene [66] is para-enriched. Also, Potemkin and Ksusainova found that water adsorbed on biological (DNA, lysozyme, collagen) and inorganic substances ($MgSO_4$, CaO, zeolites) is enriched in its para-form [69]. An explanation for this preferential adsorption of para-water on various surfaces, which seems independent of the adsorbing surface, might be its higher heat of adsorption compared to ortho-water. Thus Velikov et al. reported that the heats of adsorption Q of ortho- and para-water on activated carbon are equal to −38.9 and −46.8kJ/mol, respectively [70]. On molecular grounds, the preference for adsorption of the para-water has been attributed to the fact that this isomer can reach the ground rotational level with zero point rotational energy, which means that the molecule is not rotating. This makes the energy of interaction maximal, so that the para-water is more likely trapped/adsorbed in local surface potentials [71].

The picture emerging is that the first hydration molecular layer around the tested polyaminoacids is in every probability para-H_2O enriched, because of its zero rotational energy. The zero rotational energy and the resulting more stable interaction, means that more stable H-bonds are formed in the case of D-isomers. According to the experimental evidence, these bonds are more protective of the D-isomers towards degeneration to random structure, generated from acidic or thermal attack to the H-bonds stabilizing the helix.

In contrast, the L-isomers might form less stable H-bonds with the hydrating para-H_2O molecules. This, in combination with the selective interaction with the parity violation energy difference (PVED), could result in their increased dissolution, and eventual degeneration, in the apparent majority of the ortho-H_2O. In conclusion, the transition from the α-helix to the 3_{10}-helix as a function of temperature, is not probably symmetrical for the chiral pair poly-L-glutamic acid and poly-D-glutamic acid. In the case of poly-L-glutamic acid the content of the 3_{10} -helix remains constant and that of the α-helix decreases fast, while in the case of poly-D-glutamic acid both helices degenerate at an almost equal rate. Thus, the 3_{10}-helix appears more resistant to the destruction of its intrahelical bonds when L-aminoacid is compared to the D-aminoacid. This statement is equivalent to the expression that the water molecules existing in the vicinity of the L-isomer exhibit lower tendency in forming H-bonds with the aminoacid groups. Somehow similar results indicating asymmetrical stability of the poly-L-and the poly-D-glutamic acid as a function of pH, have been recently observed by other workers [4, 41]. This point might well be related to the well known—but still unexplained—persistence of the L-aminoacids in the living world.

Conclusion

Poly-Glutamic Acid (PGA) is a very diverse acidic biopolymer existing in two versions, the α-PGA and the γ- PGA. The second carboxyl group in the α-PGA, existing every four atoms of the polymeric chain, can interact with cationic surfactants $C_nH_{2n+1}N^+(CH_3)_3Br$ (CTAB for short) via the reaction -COOH + CTA^+ B^- → -COO-CTA +HB. The obtained soft hybridic materials form micelles which in X-ray diffraction show fingertips absolutely similar to the well-known MCM–type materials. Addition of a silica source, like TEOS, and burning of the organic part, leads to classical silicate MCM-type solids with semi-ordered mesoporosity (worm-like porosity) and very high specific surface area ~1000 m^2g^{-1}.

The packing of micelles in the soft materials PGA+CTAB is hexagonal with the hydrophilic head $N^+(CH_3)_3$ in the outside attracting and binding the side carboxyl groups of the biopolymer chain. The chains of poly-aminoacids tend to form helical structures in free form which are twisted together to form hierarhical structures.

In the case of hybrid PGA+CTAB, the poly-amino-chain seems to be twisted around the micelle, either in left or right hand mode, depending on the original chain as certified by circular dichroism: If the chain is left hand (L-PGA) the micelle (L-PGA+CTAB) is wrapped in the left hand direction, while if it is right hand (D-PGA) the micelle (L-PGA+CTAB) is wrapped in right hand.

The enantiomeric forms of free acids (L- and D-PGA), as well as the enantiomeric micelles (L-PGA+CTAB) and (L-PGA+CTAB), shows subtle but insisting differences in their hydrolytic fraying, probably via the sequence α-helix → 3_{10}–helix → random form. Such differences, observed by various groups, may be related to the unique, but still enigmatic, persistence of left-hand aminoacids, and as a result of the right-hand or α-helices, in our living world. The possible influence of ortho- and para- water in this distinct hydrolytic behavior remains elusive and speculative, but certainly worthwhile to be pursued.

Acknowledgments

This work was supported by projects 09SYN-42-791 NANOMgO and THALES-NANOMESO, financed by ERDF and National Founds.

References

[1] G.H. Ho, T.I. Ho, K.H. Hsieh, Y. C. Su, P. Y. Lin, J. Yang, K. H. Yang, S.-C. Yang, *Journal of the Chinese Chemical Society*, 53, 1363 (2006).

[2] E. Kodona, C. Alexopoulos, E. Panou, P.J. Pomonis, *Chem. Mater.*, 19, 1853 (2007).

[3] E. Kodona, C. Alexopoulos, E. Panou-Pomonis, P.J. Pomonis, *Journal of Colloid and Interface Science*, 319, 72 (2008).

[4] Y. Scolnik, I. Portnaya, U. Cogan, S. Tal, R. Haimovitz, M. Fridkin, A.C. Elitzur, D.W. Deamer, M. Shinitzky, *Phys. Chem. Chem. Phys.*, 8, 333 (2006).

[5] K. Hayagawa, J.C.T. Kwak, *J. Phys. Chem.,* 86, 3866 (1982); 87, 506 (1983).

[6] K. Hayagawa, J.P. Santerre, J.C.T. Kwak, *Macromolecules,* 16, 1642 (1983).

[7] E.D.Goddard, *Colloids Surf.,* 19, 301 (1986).

[8] M.Antonietti, J. Conrad, *Angew. Chem., Int. Ed. Engl.,* 33, 1869 (1994).

[9] M.Antonietti, G.C.Goltner, *Angew. Chem., Int. Ed. Engl.,* 36, 910 (1997).

[10] C.C.Pantazis, P.N. Trikalitis, P.J.Pomonis, M.J. Hudson, *Microporous Mesoporous Mater.,* 66, 37 (2003).

[11] C.C.Pantazis, P.J. Pomonis, *Chem. Mater.,* 15, 2299 (2003).

[12] C.C.Pantazis, P.N. Trikalitis, P.J. Pomonis, *J. Phys. Chem. B,* 109, 12574 (2005).

[13] J.S. Beck, J.C. Vartuli, W.J. Roth, C.T. Leonowicz, C.T. Kresge, K.D. Schmitt, T.-W. C. Chu, D.H. Olson, E.W. Sheppard, S.B. McCullen,. J.B. Higgins, J.L. Schlenker, *J. Am. Chem. Soc,.* 114, 10834 (1992).

[14] Q. Huo, D.I. Margolese, U. Ciesla, P. Feng, T.E. Gier, P. Sieger, R. Leon, P.M. Petroff, F. Schuth, G.D. Stucky, *Nature,* 368, 317 (1994).

[15] Q.Huo, D.I. Margolese, U. Ciesla, D.K. Demuth, P. Feng, T.E. Gier, P. Sieger, A. Firouzi, B.F. Chmelka, F. Schuth, G.D. Stucky, *Chem.* Mater., 6, 1176 (1994).

[16] A. Monnier, F. Schuth, Q. Huo, D. Kumar, D. Margolese, R.S. Maxwell, G.D. Stucky, M. Krishnamurty, P. Petroff, A. Firouzi, M. Janicke, B.F. Chmelka, *Science,* 261, 1299 (1993).

[17] D. Zhao, J. Feng, Q. Huo, N. Melosh, G.H. Fredrickson, B.F. Chmelka, G.D. Stucky, *Science,* 279, 548 (1998).

[18] D. Zhao, Q. Huo, J. Feng, B.F. Chmelka, G.D. Stucky, *J.Am. Chem. Soc.,* 120, 6024 (1998).

[19] A.P.Katsoulidis, D.E. Petrakis, G.S. Armatas, P.N. Trikalitis, P.J. Pomonis, *Microporous Mesoporous Mater.*, 92, 71 (2006).

[20] S.Mann, *Biomineralization;* Oxford University Press: New York, (2001).

[21] C.S.Sikes, A. Wierzdiski, in *Biomimetic Material Chemistry*; S. Mann, Ed.; VCH: New York, (1996) pp 249-278.

[22] U.S. Patent 5,942,150. U.S. Patent 6,913,707. U.S. Patent 7,060,199.

[23] U.S. Patent 20050037472.

[24] H. A.Lowenstam, S. Weiner, *Science,* 227, 51 (1985).

[25] G.H. Nancollas, in *Biomineralization: Chemical and Biochemical Perspectives*; S.Mann, J. Webb, R.J.P. Williams, Eds.; VCH: Weinheim, (1989) pp 157-187.

[26] H.A.Goldberg, K.J. Warner, M.C. Li, G.K. Hunter, *Connect.Tissue Res.,* 42, 25. (2001).

[27] L.Lopez - Carrasquero, S. Montserrat, A. Martinez de Ilarduya, S. Munoz-Guerra, *Macromolecules*, 28, 5535 (1995).

[28] G.Perez-Camero, M. Garcia-Alvarez, A. Martinez de Ilarduya, C. Fernadez, L. Campos, S. Munoz-Guerra, *Biomacromolecules,* 5, 144 (2004).

[29] M. Garcia-Alvarez, J. Alvarez, A. Alla, A. Martinez de Ilarduya, C. Herranz, S. Munoz-Guerra, *Macromol. Biosci.,* 5, 30 (2005).

[30] Asimov, *The Left Hand of Electron*, Doubleday, London, (1972).

[31] F.C. Frank, *Biochim. Biophys. Acta*, 11, 459 (1953).

[32] H.J. Morowitz, *J. Theor. Biol.*, 25, 491 (1969).

[33] M.H. Engel, S.A. Macko, *Nature,* 389, 265 (1997).

[34] W. Thiemann, *Jpn. Soc. Biol. Sci. Space,* 12, 73 (1998).

[35] R.F. Service, *Science,* 286, 1282 (1999).

[36] J.R. Cronin, S. Pizzarello, *Adv. Space Res.*, 23, 293 (1999).

[37] T. Cavalier-Smith, *J. Mol. Evol.*, 53, 555 (2001) .

[38] A.Y. Saghatelian, Y. Yokobayashi, K. Soltani, M.R. Chadiri, *Nature,* 409, 797 (2001).

[39] Z. Takats, S.C. Namita, R.G. Cooks, *Angew. Chem. Int. Ed.,* 42, 3521 (2003).

[40] M. Lahav, I. Weissbuch, E. Shavit, C. Reiner, G.J. Nicholson, V. Schurig, *Origins Life Evol. Biosphere,* 36, 151 (2006).

[41] (a) M.Shinitzky, D. Deamer, *Orig Life Evol Biosph.,* 38, 271 (2008); (b) A.Shvalb, Y Mastai, M. Shinitzky, *Chirality,* 22, 587 (2010).

[42] L. Pauling, R.B. Corey, *Arch. Biochem. Biophys.,* 65, 164 (1956) .

[43] S. Zhang, D.M. Marini, W. Hwang, S. Santoso, *Curr. Opin. Chem. Biol.* 6, 865 (2002).

[44] S.R. Haynes, S.D. Hagius, M.M. Juban, P.H. Elzer, R.P. Hammer, J. Pept. Res., 66, 333 (2005).

[45] G. Basu, A. Kitao, F. Hirata, N. Go, *J. Am. Chem. Soc.*, 116, 6307 (1994).

[46] M.L. Smythe, S.E. Huston, G.R. Marshall, *J. Am. Chem. Soc.*, 117, 5445 (1995).

[47] T. Wieprecht, O. Apostolov, M. Beyermann, J. Seelig, *J. Mol. Biol.,* 294, 785 (1999).

[48] R.D. Lins, R. Ferreira, *Quim. Nova,* 29, 997 (2006).

[49] (a) C.A. Rohl, A.J. Doig, *Protein Sci.,* 5, 1687 (1996) ; (b) J.K. Sun, A.J. Doig, *Protein Sci.,* 7, 2374 (1998).

[50] (a) F.B. Sheinerman, C.L. Brooks, *J. Am. Chem. Soc.* 117,10098 (1995); (b) M.L. Smythe, S.E. Huston, G.R. Marshall, *J. Am. Chem. Soc.* 115, 11594 (1993).

[51] (a) J. Tirado-Rives, D.S. Maxwell, W.L. Jorgensen, *J. Am. Chem. Soc.,* 115, 11590 (1993); (b) G.R. Marshall, E.E. Hodgkin, D.A. Langs, G.D. Smith, J. Zabrockiand, M. Leplawy, *Proc. Natl. Acad. Sci. USA,* 87, 487 (1990). (c) G.D. Rose, P.J. Fleming, J.R. Banavar, A. Maritan, *Proc. Natl. Acad. Sci. USA*, 103, 16623 (2006); (d) A. Dehner, E. Planker, G. Gemmecker, Q.B. Broxterman, W. Bisson, F. Fromaggio, M. Crisma, C. Toniolo, H. Kessler, *J. Am. Chem. Soc.,* 123, 6678 (2001). (e) M. Bellanda, S. Mammi, S. Geremia, N. Demitri, L. Randaccio, Q.B. Broxterman, B. Kaptein, P. Pengo, L. Pasquato, P. Scrimin, *Chem. Eur. J.,* 13, 407 (2007).

[52] G. Nemethy, D.C. Phillips, S.J. Leach, H.A. Scheraga, *Nature,* 214, 363 (1967).

[53] (a) E.N. Baker, R.E. Hubbard, *Prog. Biophys. Mol. Biol.,* 44, 97 (1984); (b) D.J. Barlow, J.M. Thorton, *J. Mol. Biol.,* 201, 601 (1988).

[54] T. Iwata, S. Lee, O. Oishi, H. Aoyaki, M. Ohno, K. Anzai, Y. Kirino, G. Sugihava, *J. Biol. Chem.,* 269, 4928 (1994).

[55] S. Zikou, A.-E. Koukkou, P. Mastora, M. Sakarellos-Daitsiotis, C. Sakarellos, C. Drainas, E. Panou-Pomonis, *J. Pept. Sci.,* 13, 481 (2007).

[56] M.C. Manning, R.W. Woody, *Biopolymers,* 31, 596 (1999).

[57] T.S. Yokum, T.J. Gauthier, R.P. Hammer, M.L. McLaughlin, *J. Am. Chem. Soc.*, 119, 1167 (1997).

[58] See Ref. [54].

[59] L.R. McLean, K.A. Hagaman, T.J. Owen, J.L. Krstenasky, *Biochemistry,* 30, 31 (1991).

[60] G. Subramanian (Ed.), *Chiral Separation Techniques,* Willey–VCH (2007).

[61] (a) N. Banno, T. Nakanishi, M. Matsunaga, T. Asahi, T. Osaka, *J. Am. Chem. Soc.,* 126, 428 (2004); (b) T. Nakanishi, N. Yamakawa, T. Asahi, N. Shibata, B. Ohtani, T. Osaka, *Chirality,* 16, S36 (2004).

[62] M.N. Khun, *Micellar Catalysis,* CRC Press (2007).

[63] (a) E.D. Goddad, K.P. Anathapadmanabhan (Eds.), *Interactions of Surfactants with Polymers and Proteins*, CRC Press, Boca Raton, (1993); (b) D. Rozena, S.H. Gellman, *J. Am. Chem. Soc.*, 117, 2373 (1995).

[64] (a) H.R. Trayer, Y. Nozaki, J.A. Reynolds, C. Taford, *J. Biol. Chem,.* 246, 4485 (1971); (b) E. Hinsch, J.G. Boehm, S. Groeger, F. Mueller-Schloesser, K.-D. Hinsch, *Reprod. Dom. Anim.*, 38, 155 (2003).

[65] V.I. Tikhonov, A.A. Volkov*, Science,* 296, 2363 (2002).

[66] V.I. Tikhonov, A.A. Volkov, *Chem. Phys. Chem.,* 7, 1026 (2006).

[67] The discovery of violation of parity by the weak electromagnetic forces is an exciting subject with a long history. For details see C.N. Yang, T.P. Lee, Nobel Lect. Phys. (1957); C.S. Wu et al., *Phys. Rev.* 105, 1413 (1957), describing the famous experiment which proved the parity violation hypothesis.

[68] S.F. Mason, G.E. Tranter, *Mol. Phys.,* 53, 1091, (1984).

[69] S.A. Potemkin, R.S. Ksusainova, *Biophys. Chem.* 118, 84 (2005).

[70] A.A. Velikov, S.V. Grigoriev, A.V. Chuikin, *Russian J. Phys. Chem.*, 80, 2047 (2006).

[71] X. Xin, H. Altan, A. Saint, D. Malten, R.R. Alfano, *J. Appl. Phys.,* 100, 094905 (2006).

In: Glutamic Acid
Editors: D. Balcazar and E. Perez
ISBN: 978-1-62257-236-6

Chapter 2

Metabolism and Physiology of Glutamate in Chickens

Vishwajit S. Chowdhury and Mitsuhiro Furuse
Department of Bioresource Sciences, Faculty of Agriculture,
Kyushu University, Fukuoka, Japan

Abstract

Glutamate is one of the proteinogenic and nutritionally non-essentional amino acids, and is precursor of glutamine, arginine, proline and glutathione. Glutamate also serves as the precursor for the synthesis of the inhibitory γ-aminobutyric acid (GABA) in GABAergic neurons. Among these metabolites, synthesis of arginine and proline in chickens is different from the mammals. Chickens lack carbamyl phosphate synthetase I and have relatively little ornithine transcarbamylase activities. Chickens are also unable to synthesize ornithine from glutamate and proline. These characteristics make the absolute dietary requirement of arginine in chickens. In addition, the biosynthesis of proline from glutamate is not rapid and not supports maximum growth in chicks. On the other hand, glutamate, an excitatory amino acid, acts as a neurotransmitter in the central nervous system. It can induce neuronal activity with powerful stimulatory effects. However, the central administration of glutamate attenuates stress-induced behaviors and triggers sleep-like behavior in neonatal chicks. Furthermore, glutamate plays major roles in the development of normal synaptic connections in the brain. Glutamate concentration increased in the brain with the advancement of age in developing chicks. In conclusion, glutamate is an important metabolite which plays significant roles in the development and maintenance of peripheral and central tissues in the chicken.

Metabolism of Glutamate

Glutamic acid abbreviated as Glu or E is one of the proteinogenic and nutritionally non-essential amino acids, and its codons are GAA and GAG. Glutamic acid is called as glutamate in the carboxylate anions and salts, and glutamate is a key molecule in cellular metabolism. As shown in figure 1, glutamate can serve as a precursor or substrate molecule for many compounds such as glutamine, arginine, proline, glutathione and γ-aminobutyric acid (GABA) in mammals. However, some of the compounds have a different fate and function in avian species compared to mammalian species.

In mammals, glutamate plays an important role to dispose excess nitrogen through the reaction catalyzed by glutamate dehydrogenase. Glutamate dehydrogenase converts glutamate and NADP into ammonia, 2-oxoflutarate and NADPH. Ammonia produced is toxic in even small amounts. The ammonia must be removed from the body and is excreted predominantly as urea through the urea cycle.

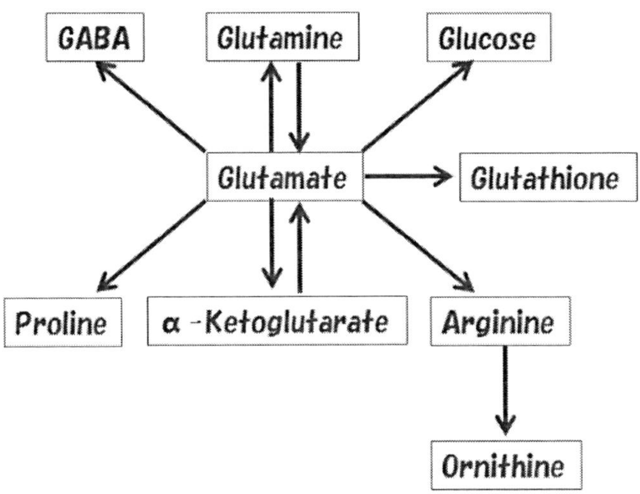

Figure 1. Metabolic pathway of glutamate.

Animals that excrete nitrogen in the form of urea are called ureotelic. The initial two steps of the urea cycle are mitochondrial as shown in figure 2. Carbamyl phosphate synthetase mediates the formation of carbamyl phosphate from NH_3^-, HCO_3^- and ATP. *N*-Acetylglutamate formed via *N*-acetylglutamate synthetase is an obligatory effector of carbamyl phosphate synthetase and an important regulator of ureagenesis. Carbamyl phosphate is converted to citrulline in the ornithine transcarbamylase reaction following condensation with ornithine. Citrulline is released to the cytosol, where it condenses with aspartate to form argininosuccinate via argininosuccinate synthetase. Argininosuccinate is cleaved in the cytosol by argininosuccinate lyase. The products of the reaction are fumarate, which is oxidized in the TCA cycle, and arginine, which is rapidly cleaved to urea and ornithine via hepatic arginase. On the other hand, birds lack carbamyl phosphate synthetase in the liver and kidney (Tamir and Ratner, 1963).

Therefore, it is impossible to synthesize arginine from glutamate in birds, and arginine is classified as an essential amino acid for birds. According to NRC (1994), arginine is required at the levels of 1.00, 0.83, 0.67 and 0.75% of the diets for 0 to 6 weeks, 6 to 12 weeks, 12 to 18 weeks and 18 weeks to first egg in the white-egg-laying strain, respectively. In the case of the brown-egg-laying strain, the values were somewhat lower, being 0.94, 0.78, 0.62 and 0.72% for each stage, respectively. Broilers, meat type chickens, require more arginine compared with laying type chickens. Arginine requirement is 1.25, 1.10 and 1.00% for 0 to 3 weeks, 3 to 6 weeks and 6 to 8 weeks, respectively.

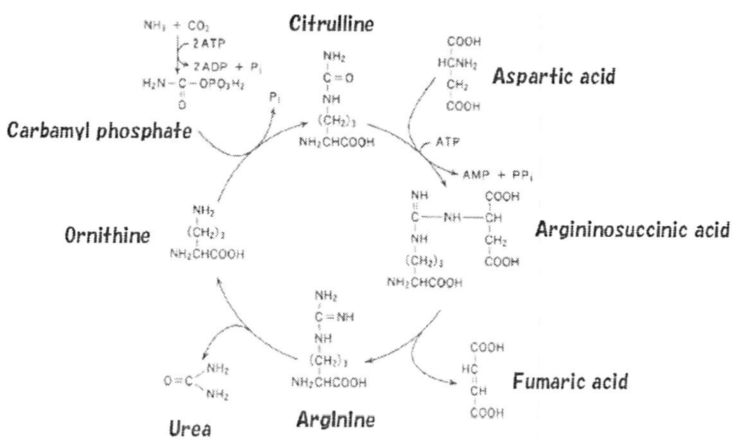

Figure 2. Urea cycle.

Arginine further metabolizes to ornithine. Ornithine is conjugated with aromatic acids such as benzoic acid to form dibenzoyl-ornithine prior to excretion through the urine in chickens (Scott et al., 1982).

In humans and higher primates, uric acid is produced as the final oxidation product of purine metabolism and is excreted in urine as the minor nitrogen compound. On the other hand, birds, terrestrial reptiles, and many insects those excrete uric acid as the major nitrogen excretory compound are called uricotelic. Xanthine oxidase makes uric acid from xanthine and hypoxanthine. The adult domestic fowl daily excretes 4 to 5 g of uric acid (Stevens, 1996). As shown in figure 3, amide nitrogen of glutamine is utilized for the uric acid synthesis. Glycine also contributes two carbon and one nitrogen atoms for the biosynthesis of uric acid. Although glycine can be synthesized by chickens, the rate is not adequate to support maximal growth (Featherston, 1976). This is due to the loss for uric acid synthesis. Serine can be converted to glycine on an equimolar basis. This reaction is reversible, and glycine can be used to form serine in chickens (Sugahara and Kandatsu, 1976). Thus, glycine and serine, which are nutritionally non-essential amino acids in mammals, are defined as nutritionally essential amino acids in chickens (NRC, 1994).

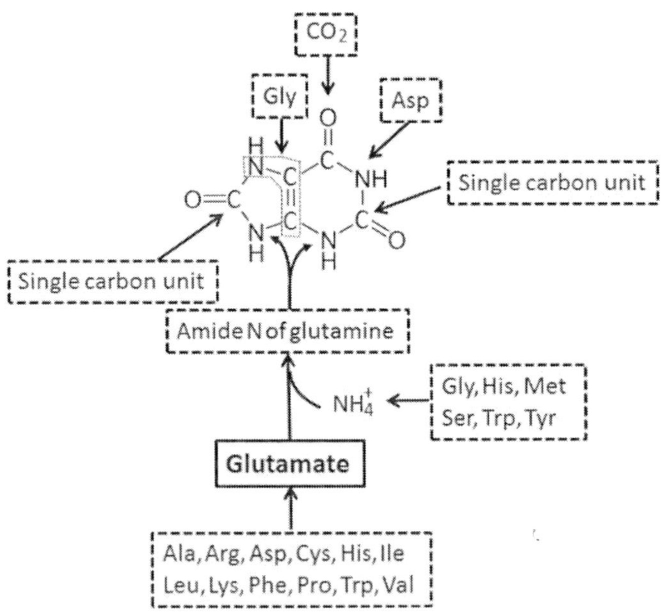

Figure 3. Components of uric acid.

Proline is formally an imino acid, but is usually treated as one of the twenty DNA-encoded α-amino acids. It is not a nutritionally essential amino acid, which means that the animal body can synthesize it from glutamate. First, the γ-carboxyl group of L-glutamate reacts with ATP to form an acyl phosphate. This mixed anhydride is then reduced by NADPH to an aldehyde. Glutamic γ-semialdehyde forms a ring structure with a loss of H_2O in a nonenzymatic process to give Δ^1-pyrroline-5-carboxylate, which is reduced by NADPH to proline. Although proline is not usually considered to be an essential amino acid for chickens, but young chicks may not synthesize sufficient proline to meet their requirements (Greene et al., 1962; Graber et al., 1970). However, requirement of proline in chickens is set only in broiler (NRC, 1994). No information is available for proline requirement in layer chickens. Glutathione is a tripeptide which consists of glutamate, cysteine and glycine. Glutathione reductase irreversibly catalyzes the reduction from oxidized glutathione to reduced glutathione (Dringen, 2000), and oxidized glutathione is a dimer of reduced glutathione. Glutathione serves to protect the brain against oxidative stress by interacting directly with reactive oxygen species or by participating in enzyme-catalyzed redox cycling reactions (Cooper and Kristal, 1997). Glutathione acts not only as an antioxidant or disulfide-reducing agent, but also as a neurotransmitter and neuromodulator in the central nervous system (CNS) (Janáky et al., 1999).

The principal precursor for GABA production is glucose *in vivo*, although pyruvate and other amino acids also can act as precursors. The first step is the transamination of α-ketoglutarate, formed from glucose metabolism in the TCA cycle by GABA α-oxoglutarate transaminase into glutamate. Glutamate decarboxylase catalyzes the decarboxylation of glutamate to form GABA.

Glutamate and aspartate, which serve as excitatory neurotransmitters, play major roles in the development of normal synaptic connections in the brain. On the other hand, GABA and glycine act as major inhibitory neurotransmitters in the brain. Lundgren et al. (1995) reported that the development of GABAergic neurons in the prosencephalon and telencephalon of chicken embryos on days 4–14 (E4–E14) is rapid; being abundant at E8, and "overexpressed" during E10–E11. This is due to the high activity of the GABA-synthesizing enzyme glutamate decarboxylase, since its activity increased approximately 25-fold from E3 to E17 (Ahman et al., 1996). Ahman et al. (1996) showed that GABA indeed accumulates during embryogenesis, whereas the levels of glutamate, the substrate for glutamate decarboxylase, were more or less unchanged up to later developmental stages. Sato et al. (2009) investigated the changes in the concentrations of excitatory and inhibitory amino acids in the brain and muscle of layer and broiler chicken embryos. Aspartate was higher in the brain than muscle, and dramatically increased in the brain at hatch. The concentration in muscle slowly increased in broilers with age, but the reverse was true in layers. GABA rapidly increased with age in the brain. Glutamate increased in the brain with age, but the maximum increase was reached in the muscle at E18. Glycine in the brain increased slowly with age. Furthermore, glycine concentration in the brain was lower than that in the muscle during embryogenesis. Glycine decreased at hatch with strong reduction in broilers. Accordingly, the concentrations of excitatory and inhibitory amino acids were different in the brain and muscle and/or in the layer and broiler chicken embryos. Hamasu et al. (2009) reported the central nervous system function of amino acids during acute stress in neonatal chicks. During the restraint with isolation-induced stress, glutamate was significantly higher, whereas GABA, arginine, and proline were significantly lower than those of the control group in the telencephalon. In the diencephalon, alanine, arginine, asparagine, aspartic acid, phenylalanine, proline, serine, and tyrosine in the restraint with isolation-induced stress group were significantly lower than those of the control group. During fasting stress, tyrosine and valine were significantly increased at 6 h after fasting, whereas arginine, histidine, and proline were significantly decreased after 3 h fasting compared with the control group in the telencephalon. In the diencephalon, tyrosine was significantly increased at 6 h fasting, whereas arginine, asparagine, aspartic acid, glutamine, histidine, phenylalanine, and proline were significantly decreased after 3 h fasting compared to the control group. It is clear that glutamate itself and the metabolites of glutamate functions in the brain during acute stress in chicks. Accordingly, central functions of glutamate and metabolites are discussed in the following sections.

Central Functions of Glutamate and Its Metabolites

Glutamate

Essential amino acids (EAAs), including glutamate and aspartate, act as neurotransmitters in the CNS. In neuroscience, glutamate is an important neurotransmitter that plays a key role in long-term potentiation and is important for learning and memory. They can induce neuronal activity with powerful stimulatory effects (Monaghan et al., 1989). Glutamate receptors are divided into two groups: iGluRs and mGluRs. The iGluRs, which form ion channels, include three main classes: α-amino-3-hydroxy-5-methyl-4-isoxazolepropionate (AMPA), kainate receptors and *N*-methyl-*D*-aspartate (NMDA). The NMDA receptor is usually blocked by Mg^{2+}. When the synaptic membrane is slightly depolarized, e.g. by previous activation of AMPA and kainate receptors, the Mg^{2+} block of the NMDA receptor is removed. The NMDA receptor is activated after binding with an agonist (Bleich et al., 2003). On the other hand, the mGluRs are coupled to GTP-binding proteins, and regulate the production of intracellular messengers. The mGluRs have eight members (mGluR1-8), categorized into three groups based on sequence homology, second messenger coupling and pharmacology. Group I includes mGluR1 and 5, group II is mGluR2 and 3 and group III is mGluR4, 6, 7 and 8. Group I mGluRs are localized postsynaptically and predominantly activate phospholipase C. Group II and III mGluRs are localized in presynaptic densities and inhibit adenylyl cyclase activity. They contribute to the regulation of synaptic plasticity and transmission (De Blasi et al., 2001; Kew and Kemp, 2005).

The distribution of iGluRs ligand binding sites is markedly different in the chick brain as shown by autoradiography (Henley et al., 1989). Although [^{3}H]L-Glutamate, [^{3}H]AMPA and [^{3}H]kainate strongly label the molecular layer of the cerebellum, the density of [^{3}H]kainate is particularly intense. [^{3}H]L-Glutamate densely labels the telencephalon, particularly the neostriatum. [^{3}H]AMPA binding sites are densely located in the hippocampus, and are also extensively distributed in the telencephalon. These regions correspond with brain structures involved in the regulation of stress, including anxiety and fear (Camargo, 2001; LeDoux, 1998; McNaughton, 1997). In general, the NMDA receptor usually coexists with the AMPA receptor in the postsynaptic membrane (Nadler, 2007). In addition, the structure and function of mGluRs has been demonstrated in several experiments in chicks (Hyson, 1998; Salinska, 2006; Zirpel et al., 1995).

Yamane et al. (2009b) reported that distress vocalizations were dose-dependently decreased by i.c.v. glutamate. The time of active wakefulness decreased dose-dependently. On the other hand, the time of sitting motionless with head drooped increased gradually. NMDA also dose-dependently affected distress vocalizations. The time of active wakefulness decreased dose-dependently. In contrast, the time of standing/sitting motionless with eyes open increased dose-dependently. AMPA tended to decrease distress vocalizations dose-dependently but this effect was not significant. The time of active wakefulness decreased dose-dependently. The behavior of standing motionless with eyes closed and sleeping posture were found only in the 150 pmol AMPA group, and the time of sleeping posture exhibited a significant increase. On the other hand, no significant effects were found by kainite in total distress vocalizations. This observation for distress vocalization by kinate was the case for behavioral changes. I.c.v. kinate also did not influence behavioral observations. Yamane et al. (2009b) also investigated the contribution of group I mGluRs, since group I mGluRs have an excitatory neurotransmission similar to iGluRs. For this objective, the group I selective mGluRs agonist (S)-3, 5-dehydroxyphenylglycine (DHPG) was applied. I.c.v. DHPG did not influence distress vocalizations and chicks receiving DHPG spent the most time in active wakefulness, and they were comparable to the control group. Standing motionless with eyes closed and sleeping posture were not found in these groups.

The i.c.v. injection of glutamate, NMDA and AMPA attenuated total distress vocalizations and induced sedation (Yamane et al., 2009b). In contrast, previous studies showed that localized infusion of glutamate induced vocalizations in cats (Bandler, 1982) and squirrel monkeys (Jürgens and Richter, 1986). Additionally, the injection of NMDA, AMPA and kainate into the substantia innominata/lateral preoptic area of rats dose-dependently stimulated locomotion (Shreve and Uretsky, 1988). Lateral hypothalamic injection of kainate also increased locomotor activity in rats (Stanley et al., 1993; Hettes et al., 2007). Further, when injected intracranially to restricted brainstem regions, NMDA elicited locomotion in decerebrate geese and ducks (Sholomenko et al., 1991). On the other hand, injection of glutamate into the lateral hypothalamus increased sleeping time (Stanley et al., 1993). The difference between these studies and Yamane et al. (2009b) may be due to species differences or the site of injection.

Although i.c.v. injection of the EAA glutamate, and the glutamate analogue NMDA, induced sedation in chicks, the behavioral results for the two compounds were somewhat different (Yamane et al., 2009b). NMDA induced an increase in the time spent standing/sitting motionless with eyes open while glutamate increased the time in sleeping posture. AMPA had a tendency to decrease wakefulness activity and increase sleep-like behavior. The central administration of kainate (0.05, 0.1, 0.25 and 0.5 μg) had no effect under stressful conditions in chicks (Panksepp et al., 1988). No significant effect of kainate was also observed compared with the control (Yamane et al., 2009b).

Henley et al. (1989) demonstrated the distribution of iGluRs ligand binding sites in the chick brain using autoradiography. The binding sites of NMDA and AMPA are related to the brain regions of stress response. However, kainite binding sites are mainly in the molecular layer of the cerebellum which is not generally considered as a stress-related region in the brain. In general, the NMDA receptor usually coexists with the AMPA receptor in the postsynaptic membrane (Nadler, 2007). Therefore, kainate may not be related to the stress response. In any case, kainate, in the doses used here, appears to have no sedative effect against isolation stress in chicks. It is suggested that the hypnotic effects of glutamate involved not only in the NMDA receptor, but also in the AMPA receptor. The group I selective mGluRs agonist DHPG did not affect sleeping posture (Yamane et al., 2009b). Although the effect of group II and III mGluRs has not been investigated, the data supports the theory that the iGluRs NMDA and AMPA, but not kaiante, are related to sleep-induction in neonatal chicks.

According to Yamane et al. (2009a), i.c.v. injection of both aspartate and aspargine attenuated the vocalization which normally occurs during social separation stress. Aspartate decreased the time spent in active wakefulness and induced sedation. Aspargine had a similar effect to Aspartate, although somewhat weaker. However, i.c.v. injection of aspartate and aspargine further enhanced plasma corticosterone release under social separation stress. Taken together, the i.c.v. injection of aspartate and aspargine has sedative effects under an acute stressful condition, which does not involve the hypothalamic-pituitary-adrenal axis. On the other hand, the effects of glutamate and glutamine on stress response were different from those of aspartate and aspargine. Glutamate induced a hypnotic and sedative effect as mentioned above, while glutamine caused no response (unpublished result). The reason for this difference was unclear.

In conclusion, the central administration of glutamate attenuates stress-induced behaviors and triggers sleep-like behavior in neonatal chicks. Furthermore, it was demonstrated that NMDA and AMPA, but not kainate and DHPG, also have a sedative effect. It is suggested that glutamate-induced hypnosis and sedation may require an interaction between NMDA and AMPA receptors. Therefore, further investigation, for example co-administration with NMDA or AMPA antagonists, is needed for a better understanding of the mechanism of glutamate. In addition, it should be confirmed in future whether NMDA and AMPA which have a sedative effect may also have an antianxiety effect or not.

Arginine and Ornithine

As mentioned above, arginine is not produced from glutamate in the chicken. If adequate amounts of arginine are produced, brain arginine may not be decreased under an acute stress as observed by Hamasu et al. (2009). However, this reduction in arginine suggests that arginine has an important role to reduce stress response. Under social separation stress, the i.c.v. injection of arginine clearly attenuated spontaneous activity and the number of vocalizations compared with the control (Suenaga et al., 2008a). In addition, arginine increased the time spent in sleeping posture. These results suggest that arginine has sedative and hypnotic effects. In their preliminary experiments, behavioral tests conducted immediately after the i.c.v. injection of arginine found no significant effects of arginine. These results suggested that a metabolite of arginine, rather than arginine itself, might participate in the central function. The i.c.v. injection of ornithine attenuated spontaneous activity, the number of vocalizations, and slightly increased the time in sleeping posture (Suenaga et al., 2008b). The effect of ornithine immediately after i.c.v. injection was stronger at 5 min of i.c.v. injection. These results indicate that ornithine was rapidly metabolized to the next metabolite. Accordingly, ornithine itself may have an important role for the induction of sedative and hypnotic effects.

Proline

The i.c.v. injection of proline examined whether centrally administered proline could modify the behavior of neonatal chicks under isolation-induced stress (Hamasu et al., 2009). When the effect of i.c.v. injection of several doses of proline on spontaneous activity and distress vocalizations during the 10 min isolation-induced stress was determined, negative correlations between the dose of proline and spontaneous activity and total distress vocalization were detected. The time for active wakefulness was reduced with increasing dose of proline. Negative correlations between the dose of proline and the time for active wakefulness were detected. Additionally, proline increased the time for standing or sitting motionless with eyes open and sleeping posture. These results revealed the novel function that proline has sedative and hypnotic effects under an acute stressful condition in neonatal chicks.

These effects of i.c.v. proline may have a considerable link to several mechanisms of neurotransmission in the CNS. Proline has been suggested as a candidate for neuronal modulator or transmitter in the CNS for many years (Snyder et al., 1973). In addition, Henzi et al. (1992) revealed that proline binds to glycine and glutamate receptors. Glycine (Asechi et al., 2006) and glutamate (Yamane et al., 2009b) cause a sedative and hypnotic effect, as observed with proline. So, the sedative and hypnotic effects of proline were possibly associated with the modulation of functions of these receptors in the brain. These possibilities are clarified later. Secondly, metabolites of proline might be responsible for sedative and hypnotic effects. Proline metabolism involves two other amino acids, glutamate and ornithine. This tri-amino acid system also links with three other essential metabolic systems, the TCA cycle, urea cycle, and pentose phosphate pathway (Hu et al., 2008). Namely, proline and its metabolites might affect this pathway and induce sedative and hypnotic effects. In conclusion, proline was significantly decreased in the telencephalon and diencephalon of neonatal chicks when several types of stress were applied. Furthermore, the i.c.v. injection of proline had sedative and hypnotic effects under an acute stressful condition in neonatal chicks. From these results, it is expected that proline might improve anxiety disorders or sleep disorders induced by a stressor.

Hamasu et al. (2010) further investigated the central effects of L-proline, D-proline and *trans*-4-hydroxy-L-proline on the acute stressful model with neonatal chicks. Sedative and hypnotic effects were induced by all compounds, while plasma corticosterone release under isolation stress was only attenuated by L-proline. To clarify the mechanism by which L-proline and D-proline induce sedative and hypnotic effects, the contribution of the strychnine-sensitive glycine receptor (glycine receptor) and NMDA receptor were investigated. The suppression of isolation-induced stress behavior by D-proline was attenuated by strychnine. However, the suppression of stress behavior by L-proline was not attenuated. The NMDA receptor antagonist (+)-MK-801 was co-injected i.c.v. with L-proline. The suppression of stress behavior by L-proline was attenuated by (+)-MK-801. These results indicate that L-proline and D-proline differentially induce sedative and hypnotic effects through NMDA and glycine receptors, respectively.

Glutathione

Hovatta et al. (2005) reported anxiety regulation by glyoxalase 1 and glutathione reductase 1 in mice. Glyoxalase 1 detoxifies dicarbonyl metabolites, and then uses glutathione as a co-factor (Thornalley, 2006).

Glutathione reductase irreversibly catalyzes the reduction from oxidized glutathione to reduced glutathione (Dringen, 2000). Reduced glutathione is a tripeptide which consists of glutamate, cysteine and glycine, and oxidized glutathione is a dimer of reduced glutathione. Reduced glutathione serves to protect the brain against oxidative stress by interacting directly with reactive oxygen species or by participating in enzyme-catalyzed redox cycling reactions (Cooper and Kristal, 1997). Glutathione acts not only as an antioxidant or disulfide-reducing agent, but also as a neurotransmitter and neuromodulator in the CNS (Janáky et al., 1999). Reduced glutathione and oxidized glutathione are endogenous ligands of the NMDA receptor, binding preferably to the glutamate recognition site via their γglutamyl moiety (Oja et al., 2000). These facts imply that central glutathione or its associated metabolites may be involved in the regulation of stress response. Greater than 99.5% of tissue "total glutathione" (i.e., reduced glutathione and oxidized glutathione, in reduced glutathione equivalents) is in the form of reduced glutathione (Anderson, 1985). Reduced glutathione is a major protectant in the brain against oxidative stress by interacting directly with reactive oxygen species or by participating in enzyme-catalyzed redox cycling reactions (Cooper and Kristal, 1997). In the reaction of the latter, glutathione peroxidase catalyzes the reduction of potentially toxic hydrogen peroxide (H_2O_2) (or lipid peroxides; ROOH) to H_2O (or ROH) with the concomitant conversion of reduced glutathione to oxidized glutathione.

Yamane et al. (2007) investigated whether i.c.v. injection of glutathione influenced isolation-induced stress responses in neonatal chicks. First, behavioral responses after i.c.v. injection of reduced glutathione and oxidized glutathione were monitored. The response to s-methylglutathione was also investigated. s-Methylglutathione is found in the brain extracellular space, and is released upon hypoxia and depolarization (Kanazawa et al., 1965). s-Methylglutathione has a methylated SH group on the cysteine moiety of reduced glutathione, and is not able to form a dimer, i.e., oxidized glutathione. Accordingly, s-methylglutathione was used to clarify the effect of reduced glutathione without the conversion to oxidized glutathione. Centrally reduced glutathione dose-dependently ameliorated distress vocalizations induced by isolation. The time for active wakefulness was reduced with increasing reduced glutathione. Additionally, the time spent in sleeping posture dose-dependently increased. Oxidized glutathione also dose-dependently suppressed vocalizations. The time for active wakefulness was reduced and sleeping posture dose-dependently increased with increasing doses of oxidized glutathione. Similarly, s-methylglutathione dose-dependently suppressed vocalization. The time for active wakefulness was reduced with increasing s-methylglutathione. When behavioral categories were divided into active wakefulness and sedation (standing/sitting motionless with eyes opened, standing motionless with eyes closed and sleeping posture), the effect of s-methylglutathione on the latter was dose-dependent.

A sleep-promoting substance was isolated from the brains of 24 h sleep-deprived rats, and one component of the sleep-promoting substance was oxidized glutathione (Komoda et al., 1990). When infused i.c.v. for 10 h during the nocturnal period, oxidized glutathione enhanced rapid-eye-movement sleep and non- rapid-eye-movement sleep in unrestrained rats (Honda et al., 1994). This result indicated that oxidized glutathione had the function of hypnosis. Honda et al. (2000) reported that oxidized glutathione was actually responsible for the reduced glutathione-induced enhancement of sleep. However, reduced glutathione had the function of hypnosis through reduced glutathione itself and/or through oxidized glutathione produced from reduced glutathione in chicks. Since reduced glutathione is structurally only half of oxidized glutathione, it is reasonable that a 2-fold greater dose is required for the somnogenicity of reduced glutathione equivalent to that of oxidized glutathione (Inoué et al., 1995). Thus, Yamane et al. (2007) applied the 2-fold dose in the reduced glutathione treatment. The obtained results were consistent with the previous report.

In conclusion, the i.c.v. injection of reduced glutathione, oxidized glutathione, and s-methylglutathione attenuates stress-induced behaviors and triggers sleeping-like behavior in neonatal chicks. There is a possibility that glutathione might be effective for the improvement of stressful insomnia induced by psychological stressors.

GABA

Watson et al. (1999) investigated the role of benzodiazepine receptors in modulating social separation-induced distress vocalizations and stress-induced analgesia in 8-day-old chicks. The results suggested that anxiolytic effects of the benzodiazepine agonist chlordiazepoxide are mediated by benzodiazepine receptor activity in the chick social separation procedure. Changes in the membrane-organization towards more fluid states favoured the accessibility of benzodiazepine to the central benzodiazepine receptor, expressed by the higher values of B(max) found in stressed samples at low temperatures with respect to control samples in forebrain membranes from chicks (Garcia et al., 2002).

The GABA-A receptor is involved in memory formation and in stress adaptive responses in early stages of chicks (Salvatierra et al., 1997). Cid et al. (2008) reported that interactions between acute stress and systemic insulin and epinephrine on GABA-A receptor density in the forebrain of 10 day-old chicks. The systemic epinephrine, perhaps by evoking central norepinephrine release, modulates the increase in forebrain GABA-A receptor binding induced by both insulin and stress.

Some of the nutrients involve attenuating stress response through the activation of GABA receptors. Takagi et al. (2003) reported that L-pipecolic acid, a major metabolic intermediate of L-lysine in the mammalian and chicken brain, activated both GABA-A and GABA-B receptors. I.c.v. injection of L-ornithine has been shown to have sedative and hypnotic effects in neonatal chicks exposed to acute stressful conditions (Suenaga et al., 2008a). To clarify the mechanism, Kurata et al. (2011) conducted experiments under strengthened stressful conditions with corticotropin-releasing factor. The chicks were injected i.c.v. with either corticotropin-releasing factor, corticotropin-releasing factor plus L-ornithine (0.5 μmol), corticotropin-releasing factor plus the GABA-A receptor antagonist picrotoxin or L-ornithine with picrotoxin. The sedative and hypnotic effects induced by L-ornithine were blocked with co-administration of picrotoxin. These results suggest that L-ornithine could attenuate corticotropin-releasing factor-stimulated stress behaviors acting on GABA-A receptors. I.c.v. injection of L-serine has been shown to have sedative and hypnotic effects on neonatal chicks exposed to acute stressful conditions (Asechi et al., 2006, 2008). Shigemi et al. (2008) studied to clarify the mechanism of function by L-serine. Co-administration of picrotoxin attenuated the sedative and hypnotic effect of L-serine. Further, Shigemi et al. (2008) also investigated the involvement of glycine receptors since L-serine is suggested to act as the alpha-homomeric glycine receptor agonist. Glycine similarly induced sedative and hypnotic effects in chicks, but its effect was attenuated by the glycine receptor antagonist strychnine. Therefore, whether the effect of L-serine was mediated through the glycine receptor was investigated using L-serine and strychnine. The effect of L-serine was inhibited by picrotoxin, but not strychnine. It appears that L-serine induces sedative and hypnotic effects by enhancing inhibitory neurotransmission via GABA-A receptors. (-)-Epigallocatechin gallate, a flavonoid (Adachi et al., 2006) and creatine (Koga et al., 2005) also functions within the CNS to attenuate the acute stress response by acting through GABA-A receptors in chicks.

It is clear that the regulation of GABA-A receptors has an important role to attenuate stress responses in the chick.

Acknowledgments

This work was supported by a Grant-in-Aid for Scientific Research from Japan Society for the Promotion of Science (Nos. 13460118, 14206033, 16380191, 17208023, 18208023, 21380165 and 23248046) and the SKYLARK Food Science Institute.

References

Adachi, N., Tomonaga, S., Tachibana, T., Denbow, D. M. and Furuse, M. (2006) (-)-Epigallocatechin gallate attenuates acute stress responses through GABAergic system in the brain. *Eur. J. Pharmacol.*, 531:171–175.

Ahman, A. K., Wågberg, F. and Mattsson, M. O. (1996) Two glutamate decarboxylase forms corresponding to the mammalian GAD65 and GAD67 are expressed during development of the chick telencephalon. *Eur. J. Neurosci.*, 8: 2111–2117.

Anderson, M. E. (1985) Determination of glutathione and glutathione disulfide in biological samples. *Methods Enzymol.* 113:548–555.

Asechi, M., Tomonaga, S., Tachibana, T., Han, L., Hayamizu, K., Denbow, D. M. and Furuse, M. (2006) Intracerebroventricular injection of L-serine analogs and derivatives induces sedative and hypnotic effects under an acute stressful condition in neonatal chicks. *Behav. Brain. Res.*, 170:71–77.

Asechi, M., Kurauchi, I., Tomonaga, S., Yamane, H., Suenaga, R., Tsuneyoshi, Y., Denbow, D. M. and Furuse, M. (2008) Relationships between the sedative and hypnotic effects of intracerebroventricular administration of L-serine and its metabolites, pyruvate and the derivative amino acids contents in the neonatal chicks under acute stressful conditions. *Amino Acids*, 34:55–60.

Bandler, R. (1982) Induction of 'rage' following microinjections of glutamate into midbrain but not hypothalamus of cat. *Neurosci. Lett.*, 30:183–188.

Bleich, S., Römer, K., Wiltfang, J. and Kornhuber, J. (2003) Glutamate and the glutamate receptor system: a target for drug action. *Int. J. Geriatr. Psychiatry*, 18: S33–S40.

Camargo, E. E. (2001) Brain SPECT in neurology and psychiatry. *J. Nucl. Med.*, 42: 611–623.

Cid, M. P., Arce, A. and Salvatierra, N. A. (2008) Acute stress *Stress*, 11:101–107.

Cooper, A. J. L. and Kristal, B. S. (1997) Multiple roles of glutathione in the central nervous system. *Biol. Chem.*, 378: 793–802.

Dringen, R. (2000) Metabolism and functions of glutathione in brain. *Prog. Neurobiol.*, 62: 649–671.

De Blasi, A., Conn, P. J., Pin, J.-P. and Nicoletti, F. (2001) Molecular determinants of metabotropic glutamate receptor signaling. *Trends Pharmacol. Sci.*, 22:114–120.

Featherston, W. R. (1976) Glycine-serine interrelations in the chick. *Fed. Proc.*, 35:1910–1913.

García, D. A., Marin, R. H. and Perillo, M. A. (2002) Stress-induced decrement in the plasticity *Mol. Membr. Biol.*, 19:221–230.

Graber, G., Allen, N. K. and Scott, H. M. (1970) Proline essentiality and weight gain. *Poult. Sci.*, 49:692–697.

Greene, D. E., Scott, H. M. and Johnson, B. C. (1962) The role of proline and certain non-essential amino acids in chick nutrition. *Poult. Sci.*, 41:116–120.

Hamasu, K., Haraguchi, T., Kabuki, Y., Adachi, N., Tomonaga, S., Sato, H., Denbow, D. M. and Furuse, M. (2009) L-proline is a sedative regulator of acute stress in the brain of neonatal chicks. *Amino Acids*, 37:377–382.

Hamasu, K., Shigemi, K., Tsuneyoshi, Y., Yamane, H., Sato, H., Denbow, D. M. and Furuse, M. (2010) Intracerebroventricular injection of L-proline and D-proline induces sedative and hypnotic effects by different mechanisms under an acute stressful condition in chicks. *Amino Acids*, 38:57–64.

Henley, J. M., Moratallo, R., Hunt, S. P. and Barnard, E. A. (1989) Localization and quantitative autoradiography of glutamatergic ligand binding sites in chick brain. *Eur. J. Nurosci.*, 1: 516–523.

Henzi, V., Reichling, D. B., Helm, S. W. and MacDermott, A. B. (1992) L-Proline activates glutamate and glycine receptors in cultured rat dorsal horn neurons, *Mol. Pharmacol.*, 41:793–801.

Hettes, S. R., Heyming, W. and Stanley, B. G. (2007) Stimulation of lateral hypothalamic kainate receptors selectively elicits feeding behavior. *Brain Res.*, 1184:178–185.

Honda, K., Komoda, Y. and Inoué, S. (1994) Oxidized glutathione regulates physiological sleep in unrestrained rats. *Brain Res.*, 636:253–258.

Honda, K., Sagara, M., Ikeda, M. and Inoué, S. (2000) Reduced glutathione regulates sleep in unrestrained rats by producing oxidized glutathione, *Sleep Hypnosis*, 2: 26–30.

Hovatta, I., Tennant, R. S., Helton, R., Marr, R. A., Singer, O., Redwine, J. M., Ellison, J. A., Schadt, E. E., Verma, I. M., Lockhart, D. J. and Barlow, C. (2005) Glyoxalase 1 and glutathione reductase 1 regulate anxiety in mice. *Nature*, 438: 662–666.

Hu, C. A., Phang, J. M. and Valle, D. (2008) Proline metabolism in health and disease. *Amino Acids*, 35:651–652.

Hyson, R. L. (1998) Activation of metabotropic glutamate receptors is necessary for transneuronal regulation of ribosomes in chick auditory neurone. *Brain Res.*, 809: 214–220.

Inoué, S., Honda, K. and Komoda, Y. (1995) Sleep as neuronal detoxification and restitution. *Behav. Brain Res.*, 69:91–96.

Janáky, R., Ogita, K., Pasqualotto, B. A., Bains, J. S., Oja, S. S., Yoneda, Y. and Shaw, C. A. (1999) Glutathione and signal transduction in the mammalian CNS. *J. Neurochem.*, 73:889–902.

Jürgens, U. and Richter, K. (1986) Glutamate-induced vocalization in the squirrel monkey. *Brain Res.*, 373:49–358.

Kanazawa, A., Kakimoto, Y., Nakajima, T. and Sano, I. (1965) Identification of gamma-glutamylserine, gamma-glutamylalanine, gamma-glutamylvaline and S-methylglutathione of bovine brain. *Biochim. Biophys. Acta,* 111:90–95.

Kew, J. N. C. and Kemp, J. A. (2005) Ionotropic and metabotropic glutamate receptor structure and pharmacology. *Psychopharmacology* (Berl.), 179: 4–29.

Koga, Y., Takahashi, H., Oikawa, D., Tachibana, T., Denbow, D. M. and Furuse, M. (2005) Brain creatine functions to attenuate acute stress responses through gabanergic system in chicks. *Neuroscience*, 132:65–71.

Komoda, Y., Honda, K. and Inoué, S. (1990) SPS-B, a physiological sleep regulator, from the brainstems of sleep-deprived rats, identified as oxidized glutathione. *Chem. Pharm. Bull.* (Tokyo), 38: 2057–2059.

Kurata, K., Shigemi, K., Tomonaga, S., Aoki, M., Morishita, K., Denbow, D. M. and Furuse, M. (2011) L-ornithine *Neuroscience*, 172:226–231.

LeDoux, J. (1998) Fear and the brain: where have we been, and where are we going? *Biol. Psychiatry*, 44:1229–1238.

Lundgren, P., Mattsson, M. O., Johansson, L., Ottersen, O. P. and Sellström, A. (1995) Morphological and GABA-immunoreactive development of the embryonic chick telencephalon. *Int. J. Dev. Neurosci.*, 13:463–472.

McNaughton, N. (1997) Congnitive dysfunction resulting from hippocampal hyperactivity – a possible cause of anxiety disorder? *Pharmacol. Biochem. Behav.*, 56:603–611.

Monaghan, D. T., Bridges, R. J. and Cotman, C. W. (1989) The excitatory amino acid receptors: Their classes, pharmacology, and distinct properties in the function of the central nervous system. *Annu. Rev. Phermacol. Toxicol.*, 29: 65–402.

Nadler, J. V. (2007) Encyclopedia of stress. In: Fink, G., McEwen, B., Ronald de Kloet, E., Rubin, R., Chrousos, G., Steptoe, A., Rose, N., Craig, I., Feuerstein, G. (ed) Excitatory amino acids, 2nd edn. Elsevier, Australia.

National Research Council (1994) *Nutrient Requirements of Poultry,* ninth revised edition. Washington, D.C.: National Academy Press.

Oja, S. S., Janáky, R., Varga, V. and Saransaari, P. (2000) Modulation of glutamate receptor functions by glutathione. *Neurochem. Int.*, 37:299–306.

Panksepp, J., Normasell, L., Herman, B., Bishop, P. and Crepeau, L. (1988) The physiological control of mammalian vocalization. In: Newman, J. D. (ed) Neural and neurochemical control of the separation distress call, Plenum Press, New York.

Salinska, E. (2006) The role of group I metabotropic glutamate receptors in memory consolidation and reconsolidation in the passive avoidance task in 1-day-old chicks. *Neurochem. Int.*, 48:447–452.

Salvatierra, N. A., Torre, R. B. and Arce, A. (1997) Learning and novelty induced increase of central benzodiazepine *Brain Res.*, 757:79–84.

Sato, M., Tomonaga. S., Denbow, D. M. and Furuse, M. (2009) Changes in free amino *Amino Acids*, 36:303–308.

Scott, M. L., Nechei, M. C. and Young, R. J. (1982) Nutrition of the Chicken. M.L. Scott and Associates, Publishers, Ithaca, New York.

Shigemi, K., Tsuneyoshi, Y., Hamasu, K., Han, L., Hayamizu, K., Denbow, D. M. and Furuse, M. (2008) L-Serine induces sedative and hypnotic effects acting at $GABA_A$ receptors in neonatal chicks. *Eur. J. Pharmacol.*, 599:86–90.

Shreve, P. E. and Uretsky, N. J. (1988) AMPA, kainic acid, and N-methyl-D-Aspartic acid stimulate locomotor activity after injection into the substantia innominata/lateral preoptic area. *Pharmacol. Biochem. Behav.*, 34:101–106.

Sholomenko, G. N., Funk, G. D. and Steeves, J. D. (1991) Avian locomotion activated by brainstem infusion of neurotransmitter agonists and antagonists. I. Acetylcholine, excitatory amino acids and substance P. *Exp. Brain Res.*, 85:659–673.

Snyder, S. H., Young, A. B., Bennett, J. P. and Mulder, A. H. (1973) Synaptic biochemistry of amino acids. *Fed. Proc.*, 32:2039–2047.

Stanley, B. G., Ha, L. H., Spears, L. C. and Dee, II M. G. (1993) Lateral hypothalamic injections of glutamate, kainic acid, D, L-α-amino-3-hydroxy-5-methyl-isoxazole propionic acid or *N*-methyl-*D*-aspartic acid rapidly elicit intense transient eating in rats. *Brain Res.*, 613: 88–95.

Stevens, L. (1996) In: Avian Biochemistry and Molecular Biology. Cambridge University Press. Cambridge.

Suenaga, R., Tomonaga, S., Yamane, H., Kurauchi, I., Tsuneyoshi, Y., Sato, H., Denbow, D. M. and Furuse, M. (2008a) Intracerebroventricular injection of L-arginine induces sedative and hypnotic effects under an acute stress in neonatal chicks. *Amino Acids*, 35:139–146.

Suenaga, R., Yamane, H., Tomonaga, S., Asechi, M., Adachi, N., Tsuneyoshi, Y., Kurauchi, I., Sato, H., Denbow, D. M. and Furuse, M. (2008b) Central L-arginine reduced stress responses are mediated by L-ornithine in neonatal chicks. *Amino Acids*, 35:107–113.

Sugahara, M. and Kandatsu, M. (1976) Glycine-serine interconversion in the rooster. *Agric. Biol. Chem.*, 40:833–837.

Takagi, T., Bungo, T., Tachibana, T., Saito, E. S., Saito, S., Yamasaki, I., Tomonaga, S., Denbow, D. M. and Furuse, M. (2003) Intracerebroventricular administration of GABA *J. Neurosci. Res.*, 73:270–275.

Tamir, H. and Ratner, S. (1963) Enzymes of arginine metabolism in chicks. *Arch. Biochem. Biophys.*, 102: 249–258.

Thornalley, P. J. (2006) Unease on the role of glyoxalase 1 in high-anxiety-related behaviour. *Trends Mol. Med.*, 12:195–199.

Watson, G. S., Roach, J. T. and Sufka K. J. (1999) Benzodiazepine receptor function in the chick social separation-stress *Exp. Clin. Psychopharmacol.*, 7:83–89.

Yamane, H., Asechi, M., Tsuneyoshi, Y., Kurauchi, I., Denbow, D. M. and Furuse, M. (2009a) Intracerebroventricular injection of L-aspartic acid *Anim. Sci. J.*, 80:286–290.

Yamane, H., Tomonaga, S., Suenaga, R., Denbow, D. M. and Furuse, M. (2007) Intracerebroventricular injection of glutathione and its derivative induces sedative and hypnotic effects under an acute stress in neonatal chicks. *Neurosci. Lett.*, 418:87–91.

Yamane, H., Tsuneyoshi, Y., Denbow, D. M. and Furuse, M. (2009b) *N*-Methyl-*D*-aspartate and α-amino-3-hydroxy-5-methyl-4-isoxazolepropionate receptors involved in the induction of sedative effects under an acute stress in neonatal chicks. *Amino Acids*, 37:733–739. Erratum, 37:767.

Zirpel, L., Lachica, E. A. and Rubel, E. W. (1995) Activation of a metabotropic glutamate receptor increases intracellular calcium concentrations in neurons of the avian cochlear nucleus. *J. Neurosci.*, 15:214–222.

In: Glutamic Acid
Editors: D. Balcazar and E. Perez
ISBN: 978-1-62257-236-6

Chapter 3

Different Roles and Applications of Poly (Glutamic Acid) in the Biomedical Field

Magdalena Stevanović
Institute of Technical Sciences of the Serbian Academy of Sciences and Arts, Belgrade, Serbia

Abstract

Poly(glutamic acid) (PGA) is a carboxylic functionalized polypeptide that has been object of intensive research.

PGA is a hydrophilic, biodegradable, and naturally available biopolymer usually produced by various strains of Bacillus. Its biological properties such as nontoxicity, biocompatibility, and nonimmunogenicity qualify it as an important biomaterial that can be easily assimilated *in vivo* rendering it usable in the field of medicine, pharmaceutics, cosmetics, food industry and others.

Poly (glutamic acid) polymer possesses appropriate physicochemical properties that allow it to be used as a reducing agent, capping agent or for the development of polymer–drug conjugates in order to achieve selective and targeted delivery of drugs.

This review article deals with the synthesis, physiochemical properties, and various biomedical applications of poly (glutamic acid).

1. Introduction

Naturally occurring polymers have attracted considerable interest in recent years.

This interest arose as a result of an increased awareness of the environment and a desire to produce environmentally safe materials. Poly (glutamic acid) (PGA) is a hydrophilic, biodegradable, and naturally available polymer.

Its biological properties such as nontoxicity, biocompatibility, and nonimmunogenicity qualify it as an important biomaterial for applications in the medicine, pharmaceutics, cosmetics, food industry and others.

Amino acids are molecules containing an amine group, a carboxylic acid group, and a side-chain that is specific to each amino acid. Amino acids can be linked together in varying sequences to form a vast variety of proteins. Twenty amino acids are naturally incorporated into polypeptides and are called proteinogenic or standard amino acids.

There are two types of amino acid: those that can be synthesized by the body itself and those that are necessary for healthy growth or maintenance but cannot be produced by the body itself.

The first group is called essential amino acids and histidine, isoleucine, leucine, lysine, methionine, phenylalanine, threonine, tryptophan and valine belong to this group.

Glutamic acid belongs to the second group, called non-essential amino acids, together with alanine, arginine, asparagines, aspartic acid, cysteine, glutamine, glycine, praline, serine and tyrosine.

Polyglutamic acid is polymerized from glutamic acid. Great insight into the formation of PGA has been gained thanks to the development of molecular, biological and physico-chemical techniques.

Moreover, there is a great variety of applications for both isoforms of PGA, many of which have not been discovered until recently.

These applications include: wastewater treatment, food products, drug delivery, biomedical materials, vaccines, using as an immune-stimulating and anti-tumor agent, etc.

This review article provides an insight about synthesis, physiochemical properties, and various biomedical applications of poly (glutamic acid).

2. Synthesis and Physiochemical Properties of Poly (Glutamic Acid)

2.1. Synthesis

PGA is a kind of polypeptide due to the repeating unit of glutamic acid, which connects α-amino and γ-carboxyl residues [1].

However, it is different from most protein in terms of accomplishing with γ-combination as amino acid's polymerism body, and its molecular weight varies from ten thousand to several hundred thousand depending on the type of bacilli used [2]. Alpha-PGA is synthesized chemically, whereas γ-PGA can be produced by a number of microbial species, most prominently various Bacilli.

Poly(γ-glutamic acid) (PGA) was found as an extracellular viscous material of *Bacillus anthracis* in 1937, [3] and Fujii [4, 5] reported that the main ingredient of natto viscosity was PGA in 1963. PGA is known as a water-soluble polymer comprising the monomer of D- and L-glutamic acids, and hydrolyzed by microorganism in the environment.Bacillus is a genus of multi-celled bacteria that has many individual species, some acting as pathogens, others as benign agents within the body. They are large, rod-shaped, genetically stable and reproduce quickly. Some even produce natural antibiotics. All this, combined with their capacity to generate a high number of enzymes in a small amount of time, makes them very useful for the creation of pharmaceutical products.Bacterial fermentation is the processes whereby bacteria consume and metabolize various types of sugars to create another substance altogether as an end result. One example would be yeast fermenting beer. In the case of bacillus, this bacterium is introduced to soybeans, which are extremely high in both polymerized amino acids and glutamate. In the absence of high sugar content, the bacillus consumes much of these amino acids to survive and excretes a long chain version of them, polyglutamic acid. Such polyglutamic acid is completely safe to eat, as it is known as a key component of natto, a traditional bean-dish from Japan. The manifest stringy "slime" can be scraped from the fermented soybeans and preserved for inclusion in pharmaceutical products.Bacillus subtilis (*chungkookjang*) was isolated from the traditional Korean fermented soybean paste, *chungkookjang*, a highly salty seasoning containing abundant γ-PGA, which is popular in Andong, South Korea (Figure 1). Forming a highly mucous colony, the strain can grow on a high-saline medium containing L-glutamic acid. Bacillus subtilis is a plasmid-free γ-PGA producer [6].

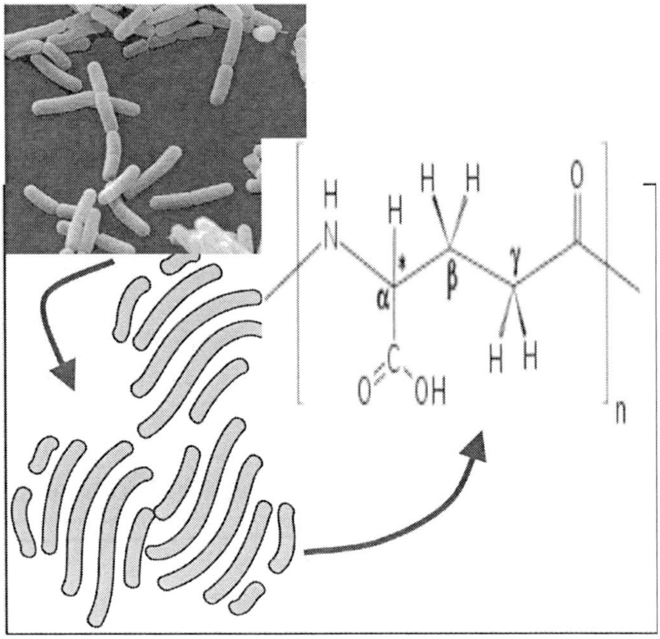

Figure 1. Poly (glutamic acid) from Bacillus subtilis.

The PGA produced by Bacillus sp. generally has high molecular weight in the range of 105–106 Da [7]. Most of the molecular weight determinations of PGA have been performed by gel permeation chromatography (GPC) using a diversity of mobile phases and calibrating against different standards [8-11].

Kino *et al.* reported poly-α-glutamic acid synthesis using a novel catalytic activity of RimK (a ribosomal protein S6-modifying enzyme) derived from *Escherichia coli* K-12 [12]. This enzyme catalyzed poly-α-glutamic acid synthesis from unprotected l-glutamic acid (Glu) by hydrolyzing ATP to ADP and phosphate. RimK synthesized poly-α-glutamic acid of various lengths; matrix-assisted laser desorption ionization-time of flight-mass spectrometry showed that a 46-mer of Glu (maximum length) was synthesized at pH 9. Interestingly, the lengths of polymers changed with changing pH. RimK also exhibited 86% activity after incubation at 55°C for 15 min, thus showing thermal stability. Furthermore, peptide elongation seemed to be catalyzed at the C terminus in a stepwise manner. Although RimK showed strict substrate specificity toward Glu, it also used, to a small extent, other amino acids as C-terminal substrates and synthesized heteropeptides. In addition, RimK-catalyzed modification of ribosomal protein S6 was confirmed. The number of Glu residues added to the protein varied with pH and was largest at pH 9.5.

2.2. Properties

Polyamino acids have various applications because of their biodegradable properties and biocompatibility. A decrease in broth viscosity of B. lichenoformis during the late stationary phase, suggesting the presence of depolymerase enzyme that breaks down the PGA have been observed by Troy et al. [13]. In the next years, a number of researchers have explored this hydrolase enzyme. The depolymerases could be either located within the cells or bound to the membrane [14]. B. subtilis IFO 3335 also expressed depolymerase activity when cultivated in a medium containing only PGA as the substrate [11]. On the other hand, nothing further has been explored on this enzyme. Goto and Kunioka [11] showed B. subtilis to grow on a membrane composed of only PGA, suggesting the presence of depolymerase enzyme. Observation that PGA concentration decrease in batch production of PGA by B. subtilis IFO 3335 during the late stationary phase due to excretion of polyglutamyl hydrolase into the broth have been reported by Richard and Margaritis in 2006 [15]. PGA was depolymerized and consumed by the microorganisms when other carbon and nitrogen sources were depleted. The cell-free broth containing PGA and polyglutamyl hydrolase enzymes was used for in situ depolymerization experiments. The temperature of 30–45°C and pH 7.0–10.0 were optimal conditions. The depolymerization reaction rate constant, k, was optimally determined to be 6.92 x 10^{-6}/h. Additional work is necessary to understand the interaction between PGA and any bound enzymes.

3. Various Biomedical Applications of Poly (Glutamic Acid)

3.1. Poly(Glutamic Acid)-Drug Carriers

Incorporation of PGA into an antigenic formulation, influenza vaccine, and gene transfection carrier has led to improved pharmaceutical efficacy. The advantage of polymeric drug carriers lies in the uptake of the polymer nanoparticles by cancer cells before they release the drug, thereby reducing its toxic effects on healthy cells. The synthesis of poly(γ-glutamic acid)-b-poly(epsilon-caprolactone)-b-poly(γ-glutamic acid) block copolymer for encapsulation anti-cancer drug doxorubicin in the treatment of wild type human breast cancer cells (MCF-7/WT) have been described in the study of

Chan and coworkers [16]. This pH-controllable carrier is negatively-charged in the presence of healthy tissues leading to lower cellular uptake. On the other hand, it becomes more hydrophobic in the acidic environment of cancer tissues, increasing its cellular uptake through the lipid bilayer. The role of poly(γ-glutamic acid) was to increase the hydrophilicity and decrease the particle size of the copolymer. The structures of micelles that were more compact and less anionic showed better stability in plasma. It was found that the drug loading content and drug loading efficiency were 12.14% and 97.22% respectively. The copolymer showed shrinking and aggregation at low pH which led to a slower drug release. These nano-sized micelles showed potential as effective drug delivery carriers for doxorubicin because of its accumulation and slow release inside the MCF-7/WT cells.

Hashida et al. used α-PGA as a polymeric backbone and galactose moiety as a ligand to target hepatocytes. Their in vivo results indicated that the galactosylated α-PGA had a remarkable targeting ability to hepatocytes and degradation of α-PGA was observed in the liver [17].

A new series of PGA based carriers as suitable vectors for DNA have been developed by Dekie et al. [18]. The complex between DNA and PGA was fairly stable towards serum albumin. Preliminary transfection data exhibited that the complexes were able to transfect spontaneously 293 cells. PGA derivatives readily form polyelectrolyte complexes with DNA, resulting in a reduced surface charge and size of the DNA. Such complexes are able to protect DNA from digestion by DNase I, and are also capable of transfecting cells. Nanoparticles of PGA are superior gene carrier in future clinical applications because of their DNA protection and gene-transfer efficiency.

Fante et al. report the synthesis of a polyglutamic acid-dopamine conjugate and show that conjugation significantly extends (from 1 to 24 h) dopamine's antiangiogenic activity in vitro and in vivo [19]. Dopamine has previously been shown to inhibit angiogenesis in vitro and in vivo, but its clinical applications in this context are severely limited by its short half-life. The results from this study form the basis for the development of a new class of agents for the treatment of angiogenesis-dependent diseases.

Poly(glutamic acid) nanoparticles are excellent delivery carriers for tumor vaccines. PGA can deliver antigenic proteins to antigen-presenting cells (APCs) and eliciting potent immune responses based on antigen-specific cytotoxic T lymphocytes [20]. Antigen uptake by APCs is also enhanced by the association of the antigens with polymeric micro/nanoparticles. Notably, PGA nanoparticles efficiently deliver entrapped antigenic proteins through cytosolic translocation from the endosomes, which was a key process of PGA nanoparticles -mediated anti-tumor immune responses. This suggests the suitability of the PGA system for the intracellular delivery of protein-based drugs as well as tumor vaccines. Technique to prepare uniform nanoparticles using PGA, in which L-phenylalanine ethyl ester (LPAE) was introduced as a hydrophobic residue into the a-position group carboxyl of PGA have been developed by Nakagawa et al. [20]. In mice, subcutaneous immunization with PGA nanoparticles entrapping ovalbumin more effectively inhibited the growth of ovalbumin -transfected tumors than immunization with ovalbumin emulsified using CFA. In addition, PGA nanoparticles did not induce histopathologic changes after subcutaneous injection or acute toxicity through intravenous injection. The immunization with ovalbumin carrying PGA nanoparticles to induce significant expansion of antigen-specific CD^{8+} T cells has been demonstrated by Uto et al. [21]. Dissimilar CFA, subcutaneous inoculation of ovalbumin nanoparticles to footpad did not generate swelling at injection site. Although these nanoparticles could induce both antigen-specific cellular and humoral immune responses,the dominant induction of either cellular or humoral immunity depend on how they are administrated. Strong antibody production was observed by subcutaneous immunization, yet no antibody. Okamoto et al. [22]. showed a single dose of JE vaccine with PGA-NPs to enhance the neutralizing antibody titer and all the immunized mice survived a normally lethal JEV infection, while only 50% of the mice that received a single dose of JE vaccine without PGA-NPs could survive.

Amphiphilic poly(γ-glutamic acid)-*graft*-L-phenylalanine copolymers have been used to prepare biodegradable nanoparticles showing great potential as drug and protein carriers [23-25].

These and other nanoparticles made of analogous aminoacid-grafted PGGA have been reported to be efficient as antigen carriers.The manufacture of polymer particles for protein delivery carrier systems requires special precautions because the complex structure of the protein molecule and its sensitivity to undergo conformational changes.

A paclitaxel poliglumex (PPX), a macromolecular taxane, to increase the therapeutic index of paclitaxel have been synthesized by Singer et al. [26]. This large macromolecular conjugate of paclitaxel and PGA accumulated in tumor tissues by virtue of enhanced permeability of tumor vasculature and lack of lymphatic drainage. Preclinical studies in animal tumor models demonstrated PPX to be more effective than standard paclitaxel, and were associated with prolonged tumor exposure to active drug while minimizing systemic exposure. Phase 1 and 2 clinical studies with PPX showed encouraging outcomes compared to standard taxanes with reduced neutropenia and alopecia and allowed more convenient administration schedule without need for routine premedications. Human pharmacokinetic data were consistent with prolonged tumor exposure to active drug and a limited systemic exposure. Paclitaxel–PGA complex is also reported to reduce TCD50 of single dose irradiation from 53.9 Gy to 7.5 Gy and that too without affecting the normal tissue radioresponse.

3.2. Applications of Poly (Glutamic Acid) in Cosmetics

Poly (glutamic acid) can be used as a component to enhance the properties of existing or novel personal care products as a moisturizer, exfoliant and wrinkle-remover.

The chemistry of PGA allows it to be homogenously miscible as well as chemically stable in a matrix of ingredients characteristically used in facial creams [27]. Additional qualities of PGA include the ability to form a smooth, elastic and soft film on the skin it is an excellent hydrophilic humectant and can increase the production of natural moisturizing factors as pyrrolidone carboxylic acid, lactic acid and urocanic acid.

A safe and natural composition containing PGA and extract of aloe vera or green tea has been shown to increase pyrrolidonecarboxylic acid on skin epidermis without disturbing the quantitative balance and constancy of a humectant ingredient essential to the skin [27]. PGA inhibits hyaluronidase that degrades hyaluronic acid (which is a element of skin connective tissue) and helps to maintain the elasticity of the skin. Microscopic assessment of the PGA hydrogel reveals a multibaglike structure that enables it to absorb moisture $5x10^3$ times its own weight.

The amount of water contained within the PGA hydrogel is influenced by pH and salt content. This allows cosmetic formulators to improve the moisturizing ability of hair or skin formulations by adding proper amount of electrolytes and appropriate pH adjustments. The release of moisture is important in applications such as facial masks that are used to hydrate the skin as well as to reduce the appearance of wrinkles [27]. PGA can increase hair strength to improved withstand the bleaching process by increasing its moisture retention ability and forming a protective barrier capable of diluting the chemical interactions of the applied colorings with the protein makeup of the hair [27].

3.3. Poly(Glutamic Acid) in Food Industry

Sakai et al. [28] showed that the addition of poly(glutamic acid) to substances having a bitter taste (amino acids, peptides, quinine, caffeine, minerals, etc.), for prevention of aging and improvement of textures of starch-based bakery products and noodles, and as an ice cream stabilizer. PGA is reported to increase bioavailability of calcium by increasing its solubility and intestinal absorption. An increased intestinal absorption of calcium even in post-menopausal women after a single dose of PGA has been demonstrated by Tanimoto et al. (2007) [29]. The increase in solubility is attributed to the inhibition of the formation of an insoluble calcium phosphate. Addition of poly(glutamic acid) in the high mineral food preparations accelerated the absorption of minerals in the small intestine and also masked the extraneous taste of enriched minerals. This effect has also been seen in animal feeds where addition of PGA expedite the absorption of minerals such as phosphorus, increases the strength of egg shells, or retains the effect of decreasing the accumulation of body fat [29].

3.4. Poly(Glutamic Acid) as Sorbent

PGA is a potential sorbent for the removal and recovery of heavy metals from industrial wastewaters due to its ability to bind several metal ions (Ni+2, Cu+2, Mn+2 and Al+3). Bhattacharyya et al. [30] developed microfiltration membranes with covalently attached PGA that have extremely high capacities for absorption of heavy metals. PGA obtained from B. licheniformis ATCC 9945 had a very good copper adsorption capacity approaching 77.9 mg/g and a binding constant of 32 mg/l at pH 4.0 and 25 _C (Mark et al., 2006) [31]. An excellent mercury binding capacity PGA has been confirmed by Inbaraj et al. [32]. The preparation of novel biodegradable nanoparticles based on complexation of PGA with bivalent lead ion has been described by Bodnar et al. [33]. The strong complexation capacity of PGA for lead ions indicated a promising sorbent for removal of heavy metals in polluted water.

3.5. Poly(Glutamic Acid) as Cryoprotectant

PGA was displayed high antifreeze activities because of its high acidic amino acid composition. Investigations on antifreeze activities of PGA from

B. licheniformis by DSC showed that PGA with molecular weight below 20,000 had antifreeze activity higher than that of glucose. The antifreeze activities of PGA were slightly affected by the optical isomerism and decreased in the order Na salt = K salt_Ca salt_acidic form. Shih et al. [34-37] demonstrated the antifreeze activity of PGA to increase with a decrease in its molecular weight, but was indifferent to its D/L-glutamate composition.

3.6. PGA as Capping Agent

PGA has been used as capping agent or stabilizer for many different colloidal particles (Figure 2). The choice of an appropriate stabilizer and reducing agent for preparation of stable colloidal silver nanoparticles is very important [43]. Poly (α, γ, L-glutamic acid) (PGA) has been used as an organic layer for silver nanoparticles, i.e., as a capping agent to make the silver nanoparticles more biocompatibile and protect them from agglomeration in the suspension medium [44]. Poly (α, γ, L-glutamic acid) is a hydrophilic, biodegradable, and naturally available biopolymer usually produced by various strains of *Bacillus*. Its biological properties such as nontoxicity, biocompatibility, and nonimmunogenicity qualify it as an important biomaterial for applications in the medicine, pharmaceutics, cosmetics, food industry and others. If, for example, silver nanoparticles are intended for use in biomedical purposes, PGA can ensure a more favorable interaction of the nanoparticles with living cells and at the same time can act as particle stabilizer [43].

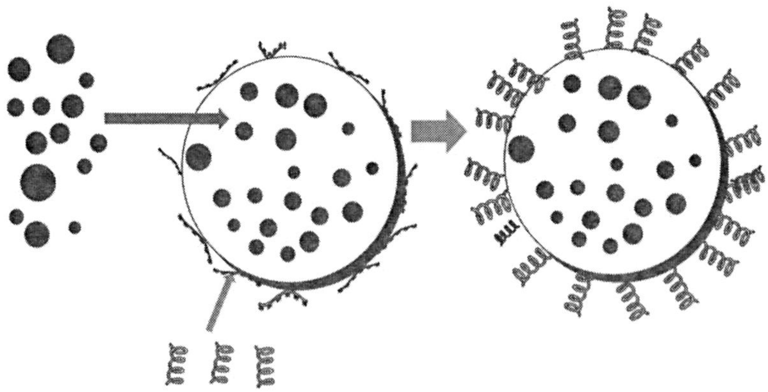

Figure 2. PGA as capping agent.

Conclusion

As discussed in this article poly (glutamic acid) has many potential applications in different biomedical fields due to its high solubility, nontoxicity, and biodegradability and considerable research on its production and application have been carried out. Until now PGA has been used as drug carrier, sorbent, capping agent, cryoprotectant and finds many applications in medicine, pharmacy, food industry, etc. New derivates of PGA have the potential to be used as new drug delivery systems, plastics, glue or vaccine adjuvant. This review will also be a contribution to the development of commercial scale production and applications of PGA.

References

[1] H. T. Zwartouw, H. Smith, Polyglutamic Acid from Bacillus anthracis Grown in vivo:Structure and Aggressin Activity, *Biochem. J.,* 1956 July; 63(3): 437–442.

[2] L.C Yang, J-B. Wu, G-H. Ho, S-C. Yang, Y-P. Huang, W-C Lin, Effects of poly-y-glutamic acid on calcium absorption in rats, Biosci. Biotechnol. *Biochem.,* 72 (12), 3084-3090, 2008.

[3] Seong-Hyun Choi, Jong-Soo Park, Kyung-Sook Whang, Min-Ho Yoon, Woo-Young Choi, Production of Microbial Biopolymer, Poly(γ-glutamic acid), by Bacillus subtilis BS 62, Agric. *Chem. Biotechnol.,* 47(2), 60-64 (2004).

[4] Ivanovics, G. and V. Bruckner. Chemische und immunologische Studien uber den Mechanismus der Milzbrandinfektion und Immuni-tat; die chemiseche Struktur der Kapselsubstanz des Milzbrand-bazillus und der serologisch identischen spezifischen Substanz des *Bacillus mesentericus. Z Immunitatsforsch,* (1937) 90, 304-318.

[5] Fujii, H. On the formation of mucilage in natto (1). *Nippon Nogeikagaku Kaishi,* (1963) 37, 407-411.

[6] Moon-Hee Sung, Chung Park, Chul-Joong Kim, Haryoung Poo, Kenji Soda, Makoto Ashiuchi, Natural and Edible Biopolymer Poly-g-glutamic Acid: Synthesis, Production, and Applications, *The Chemical Record,* (2005)Vol. 5, 352–366

[7] Shih, I.L., Van, Y.T., Chang, Y.N., Application of statistical experimental methods to optimize production of poly(c-glutamic acid)

by Bacillus licheniformis CCRC 12826. *Enzyme Microb. Technol.,* (2002) 31, 213–220.

[8] Kunioka, M., Goto, A., Biosynthesis of poly(c-glutamic acid) from L-glutamine, citric acid and ammonium sulfate in Bacillus subtilis IFO3335. Appl. Microbiol. *Biotechnol.,* (1994) 40, 867–872.

[9] Kunioka, M., Biosynthesis of poly (c-glutamic acid) from L-glutamine, citric acid and ammonium sulfate in Bacillus subtilis IFO3335. *Appl. Microbiol. Biotechnol.,* (1995)44, 501–506.

[10] Kunioka, M., Biosynthesis and chemical reactions of poly (amino acid)s from microorganisms. *Appl. Microbiol. Biotechnol.,* (1997) 47, 469–475.

[11] Goto, A., Kunioka, M., Biosynthesis and hydrolysis of poly(c-glutamic acid) from Bacillus subtilis IFO3335. *Biosci. Biotechnol. Biochem.,* (1992) 56, 1031–1035.

[12] Kuniki Kino, Toshinobu Arai and Yasuhiro Arimura, Poly-α-Glutamic Acid Synthesis Using a Novel Catalytic Activity of RimK from Escherichia coli K-12, *Appl. Environ. Microbiol.,* (2011), 77(6): 2019.

[13] Troy, F.A., Chemistry and biosynthesis of the poly(c-D-glutamyl) capsule in Bacillus licheniformis. I. Properties of the membrane-mediated biosynthetic reaction. *J. Biol. Chem.,* (1973) 248 (1), 305–315

[14] Birrer, G.A., Cromwick, A.M., Gross, R.A., Poly(glutamic acid) formation by Bacillus licheniformis 9945a: physiological and biochemical studies. *Int. J. Biol. Macromol.,* (1994) 16, 265–275.

[15] Richard, A., Margaritis, A., Empirical modeling of batch fermentation kinetics for poly(glutamic acid) production and other microbial biopolymers. *Biotechnol. Bioeng.,* (2004) 87 (4), 501–515.

[16] Chan AS, Chen CH, Huang CM, Hsieh MF. Regulation of particle morphology of pH-dependent poly(epsilon-caprolactone)-poly(gamma-glutamic acid) micellar nanoparticles to combat breast cancer cells. *J. Nanosci. Nanotechnol.,* (2010) 10(10):6283-97.

[17] Hashida, M.; Akamatsu, K.; Nishikawa, M.; Yamashita, F.; Takakura, Y. *J. Controlled Release,* (1999), *62*, 253.

[18] Do, J.H., Chang, H.N., Lee, S.Y., Dekie, L., Toncheva, V., Dubruel, P., Schacht, E.H., Barrett, L., Seymour, L.W., Poly-L-glutamic acid derivatives as vectors for gene therapy. *J. Controlled Release,* (2000) 65, 187–202.

[19] Cristina Fante, Anat Eldar-Boock, Ronit Satchi-Fainaro, Helen M. I. Osborn and Francesca Greco, Synthesis and Biological Evaluation of a Polyglutamic Acid–Dopamine Conjugate: A New Antiangiogenic Agent, *J.. Med. Chem.,* 2011, 54 (14), pp 5255–5259

[20] Nakagawa, S., Efficacy and safety of poly (c-glutamic acid) based nanoparticles (c-PGA NPs) as vaccine carrier. *Yakugaku Zasshi,* (2008) 128 (11), 1559–1565.

[21] Uto, T., Wang, X., Akagi, T., Zenkyu, R., Akashi, M., Baba, M., Improvement of adaptive immunity by antigen-carrying biodegradable nanoparticles. Biochem. Biophys. *Res. Commun.,* (2009) 379 (2), 600–604.

[22] Okamoto, S., Yoshii, H., Ishikawa, T., Akagi, T., Akashi, M., Takahashi, M., Yamanishi, K., Mori, Y., Single dose of inactivated Japanese encephalitis vaccine with poly(c-glutamic acid) nanoparticles provides effective protection from Japanese encephalitis virus. *Vaccine,* (2008) 26 (5), 589–594.

[23] T. Akagi, T. Kaneko, T. Kida, M. Akashi,Preparation and characterization of biodegradable nanoparticles based on poly(y-glutamic acid) with L-phenylalanine as a protein carrier, *J. Control. Rel.,* (2005), *108*, 226.

[24] T. Akagi, M. Higashi, T. Kaneko, T. Kida, M. Akashi, Hydrolytic and enzymatic degradation of nanoparticles based on amphiphilic poly(y-glutamic acid)-graft-L-phenylalanine copolymer, *Biomacromolecules,* (2006), 7, 297.

[25] T. Akagi, M. Baba, M. Akashi, Preparation of nanoparticles by the self-organization of polymers consisting of hydrophobic and hydrophilic segments: *potential applications Polymer,* (2007) 48: 6729-6747.

[26] Singer, J.W., Paclitaxel poliglumex (CT-2103): a macromolecular taxane. *J. Controlled Release,* (2005) 109, 120–126.

[27] Ben-Zur, N., Goldman, D.M., c-Poly glutamic acid: a novel peptide for skin care. *Cosmetics Toiletries Mag.,* (2007) 122 (4), 64–72.

[28] Sakai, K., Sonoda, C., Murase, K., 2000. Bitterness relieving agent. WO0021390.

[29] Tanimoto, H., Fox, T., Eagles, J., Satoh, H., Nozawa, H., Okiyama, A., Morinaga, Y., Susan, J., Fairweather-Tait, S.J., Acute effect of poly-glutamic acid on calcium absorption in post-menopausal women. *J. Am. College Nutr.,* (2007) 26 (6), 645–649

[30] Bhattacharyya, D., Hestekin, J.A., Brushaber, P., Cullen, L., Bachas, L.G., Sikdar, S.K., Novel polyglutamic acid functionalized microfiltration membranes for sorption of heavy metals at high capacity. *J. Membr. Sci.,* (1998) 141, 121–135.

[31] Mark, S.S., Crusberg, T.C., DaCunha, C.M., Di Iorio, A.A., A heavy metal biotrap for wastewater remediation using poly-c-glutamic acid. *Biotechnol. Prog.,* (2006) 22, 523–531.
[32] Inbaraj, B.S., Wang, J.S., Lu, J.F., Siao, F.Y., Chen, B.H., Adsorption of toxic mercury(II) by an extracellular biopolymer poly(c-glutamic acid). *Bioresour. Technol.,* (2009) 100, 200–207.
[33] Bodnar, M., Kjonikse, A.L., Molnar, R.M., Hartmann, J.F., Daroczi, L., Nystromb, B., Borbely, J., Nanoparticles formed by complexation of poly-gammaglutamic acid with lead ions. *J. Hazard. Mater.,* (2008) 153, 1185–1192.
[34] Shih, I.L., Van, Y.T., Chang, Y.N., Application of statistical experimental methods to optimize production of poly(c-glutamic acid) by Bacillus licheniformis CCRC 12826. *Enzyme Microb. Technol.,* (2002) 31, 213–220.
[35] Shih, I.L., Van, Y.T., Sau, Y.Y., Antifreeze activities of poly (c-glutamic acid) produced by Bacillus licheniformis. *Biotechnol. Lett.,* (2003) 25 (20), 1709–1712.
[36] Shih, I.L., Van, Y.T., Yeh, L.C., Lin, H.G., Chang, Y.N., Production of a biopolymer flocculant from Bacillus licheniformis and its flocculation properties. *Bioresour. Technol.,* (2001) 78, 267–272.
[37] Shih, I.L., Van, Y.T., The production of poly(c-glutamic acid) from microorganism and its various applications. *Bioresour. Technol.,* (2001) 79, 207–225.
[38] Anne-Marie Cromwick, Gregory A. Birrer and Richard A. Gross, "Effects of pH and Aeration on γ-Poly(glutamic acid) Formation by Bacillus licheniformis in Controlled Batch Fermenter Cultures", *Biotechnol. Bioeng.,* Vol. 50, 222-227 (1996).
[39] Anne-Marie Cromwick and Richard A. Gross, "Effects of Manganese (II) on γ-Poly(glutamic acid) Formation by Bacillus licheniformis ATCC 9945A, *Int. J. Biol. Macromol.,* (1995) Vol 17, No. 5, 259-267.
[40] Anne-Marie Cromwick and Richard A. Gross, "Investigation by NMR of Metabolic Routes to Bacterial γ-Poly(Glutamic Acid) Using 13C Labeled Citrate and Glutamate as Media Carbon Sources, *Can. J. Microbiol.,* (1995) Vol 41: 902-909.
[41] Gregory A. Birrer, Anne-Marie Cromwick and Richard A. Gross, "γ-Poly(glutamic acid) Formation by Bacillus licheniformis ATCC 9945A: Physiology and Biochemical Studies, *Int. J. Biol. Macromol.;* (1994)16(5) 265-275.

[42] Sakiyama, M.; Seki, S., Enthalpies of combustion of organic compounds. II. L- and D-glutamic acid, *Bull. Chem. Soc. Jpn.*, (1975), 48, 2203-2204.

[43] M. Stevanović, I. Savanović, V. Uskoković, S. D. Škapin, I. Bračko, U. Jovanović, D. Uskoković -A new, simple, green, and one-pot four-component synthesis of bare and poly(α, γ, L-glutamic acid) capped silver nanoparticles- *Colloid and Polymer Science,* (2012) 290: 221-231.

[44] M. Stevanović, B. Kovačević, J. Petković, M. Filipič, D. Uskoković, Effect of poly (α, γ, L-glutamic acid) as capping agent on the morphology and oxidative stress-dependent toxicity of silver nanoparticles, *International Journal of Nanomedicine,* (2011):6 2837–2847.

In: Glutamic Acid
Editors: D. Balcazar and E. Perez

ISBN: 978-1-62257-236-6

Chapter 4

Glutamic Acid in Food and Its Thermal Degradation in Acidic Medium

Giuseppe Montevecchi,[1,*] *Francesca Masino*[1,2] *and Andrea Antonelli*[1,2]
[1]Centro di Ricerca Interdipartimentale per il Miglioramento e la Valorizzazione delle Risorse Biologiche Agro-Alimentari BIOGEST-SITEIA, Università degli Studi di Modena e Reggio Emilia, Italy
[2]Dipartimento di Scienze Agrarie e degli Alimenti, Padiglione Besta, Università degli Studi di Modena e Reggio Emilia, Italy

Abstract

Glutamic acid is an amino acid naturally occurring in many foods and it is responsible for umami taste. For this reason, it has been widely used as a food additive and flavor enhancer as monosodium glutamate.

Glutamic acid is stable under standard conditions. However, high temperature or extreme pH conditions may induce racemization. In addition, it is a good substrate for non-enzymic browning reactions (Maillard reaction).

A case study of the glutamic acid thermal degradation at acidic pH was carried out. Different grape musts were subjected to heating at 90 °C

* Tel: +39-0522-522061, Fax: +39-0522-522027; Email: giuseppe.montevecchi@unimore.it.

for 30 h by means of a lab-scale equipment emulating a real process. Model solutions were utilized to gain a deeper comprehension of this phenomenon and to explain the glutamic acid degradation pattern. Results showed that glutamic acid underwent degradation during grape must cooking, following the same trend observed in the model solutions. The amino acid was almost linearly reduced throughout all the cooking procedure yielding pyroglutamic acid as the main product of degradation.

1. Introduction

Glutamic acid (Glu or E) is a proteinogenic non-essential amino acid naturally occurring in many foodstuffs. It can be found as protein-bound glutamate or in its free form. Seasoned cheeses (such as *Parmigiano*, *Pecorino*), legumes (such as lupins, lentils), mushrooms, tomatoes, milk, meat (such as beef, pork) and poultry, fish (such as codfish, lake whitefish), almonds and pumpkin seeds, and seaweeds (spirulina) are some of the high-Glu content foods. Table 1 gives an outlook of the Glu content in some common foodstuffs.

Glu was isolated for the first time in the wheat gluten hydrolyzate [1] and hence its name "*glutaminsäure*". Later, Glu chemical structure was identified by Wolff in 1890 [2].

In its free form, Glu has a low solubility (8.64 g/L at 25 °C) [3] and does not exhibit a defined taste, while its sodium salt is responsible for umami taste. Umami is classified as the fifth basic taste in addition to sweet, sour, salty, and bitter. This distinctive taste was first described by Ikeda [4, 5] during his study on the flavor of *kombu* (or *konbu*), a typical Eastern Asian seaweed. *Kombu* is an edible kelp from the family *Laminariaceae*. The term "umami" passed into English due to the absence of an exact synonymous [6]. The umami is due to L-form of α-amino dicarboxylates that has four to seven carbons [7, 8]. For this reason, it has been widely used as a food additive and flavor enhancer as monosodium glutamate (MSG) [9]. MSG is usually added pure or as yeast extracts or as hydrolyzed proteins containing high percentages of Glu [10, 11].

Umami is considered so likable because glutamate is present in traces in albumen food matrix and in animal foodstuff. In vegetable matrices, free Glu is even more abundant than in animal ones, but disodium salts of the 5′-nucleotides inosine monophosphate (IMP), guanosine monophosphate (GMP), and adenosine monophosphate (AMP) in animal foodstuff enhance the umami effect due to glutamate ion. That being so, it is likely that umami can represent a signal to indicate animal proteins [12, 13].

Table 1. Glutamic acid content (min and max) in foods

Food	Protein in food (%)[a]	Glutamate in protein (%)[b]	Protein-bound glutamate (g/100 g)[a]	Free glutamic acid (mg/100 g)
Milk	1.1-2.9	15.5-21.7[d]	0.170-0.560	1[g]-22[h]
Cheese	17.5[c]-36.0[c]	27.4	4.787[c]-9.847[c]	22[f]-1680[g]
Eggs	12.8	12.0[e]-12.5	1.600	23[i]
Meat and poultry	15.8[d]-22.9	13.5-17.1[e]	2.500-3.700	9[g]-44[j]
Vegetables	0.7-7.4	9.1-37.1	0.190[f]-1.510[f]	10[g]-246[k]
Seafood				20[g]-140[g]
Seaweed				9[g]-1608[g]
Fruits				5[g]-18[g]

[a] [34].,[b] [8].,[c] [35].,[d] [22].,[e] [36].,[f] [37].,[g] [6].,[h] [38].,[i] [39].,[j] [40].,[k] [41].

Even if the Food and Drug Administration includes MSG among the substances generally recognized as safe (GRAS), and there is a general consensus in the scientific community about the MSG safety [14], a demand for “natural” flavorings without glutamate and other chemical enhancers is currently growing. Many emerging spice blends or seasonings are being formulated to follow this market trend. Moreover, sustainable methods of growing crops and manufacturing products without the addition of chemicals (such as MSG, hydrolyzed plant protein, salt, sugar, or chemical preservatives) are the today trend. Organic spices, produced under eco-farming and sustainable methods, free from chemical contaminants and pesticide residues, are the obvious reply to this request [15].

Glu is not hygroscopic and it does not change its appearance under standard conditions, and it is stable under food normal processing and cooking. However, high temperature or extreme pH conditions may induce racemization, thus yielding to DL-glutamate.

Alkaline conditions are more drastic than acid ones to induce racemization. In addition, it is a good substrate for non-enzymic browning reactions (Maillard reaction). In general, this reaction occurs in the presence of reducing sugars and is boosted by alkaline pH and high temperature [8].

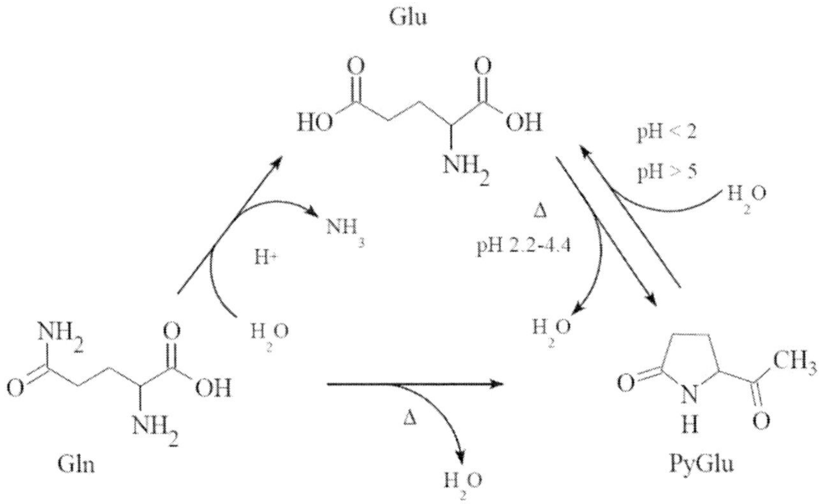

Figure 1. Glutamic acid decomposition and formation pathways. Glu: glutamic acid; Gln: glutamine; Pyglu: pyroglutamic acid.

Nunes and Cavalheiro [16] submitted Glu to thermal analysis and they verified that pyroglutamic acid (also known as 5-oxoproline, pidolic acid, 5-oxopyrrolidine-2-carboxylic acid) was the main decomposition product of the first thermal event collected at 200 °C. Through a water loss (1 mole of water/mol Glu) Glu cyclizes to yield pyroglutamic acid. In other words, the dehydration and cyclization converts Glu into its lactam, pyroglutamic acid.

After the formation of pyroglutamic acid, Glu decomposition takes place in two consecutive steps, without any residue at 700 °C [16]. However, these latter reactions are not of food concern.

Glutamine could yields pyroglutamic acid [17, 18], as well. When heated, in fact, glutamine releases ammonia on a molar basis, mainly due to the deamination of the amide group and pyroglutamic acid [19], contrary to more stable lower homologue, asparagine [20] (Figure 1).

Also under acidic (pH 2.2–4.4) conditions together with high temperatures, a portion of Glu is dehydrated and converted into pyroglutamic acid [21], while the latter is cleaved again to Glu at different pHs [22]. Pyroglutamic acid has a sour taste and does not show any toxic effect when orally administered at 1-5 g/day [23].

A case study of the Glu thermal degradation at acidic pH was described. Different grape musts were subjected to heat-concentration at 90 °C for 30 h by means of a lab-scale equipment emulating the real process used to obtain cooked must [24]. Cooked must is a product largely diffused in many wine-producing countries, where it is used as it is, for typical local recipes, as for some Spanish sweet wine [25] or for the production of some typical Italian vinegars and wines or as an ingredient in baking [26, 27].

The degradation trends observed for Glu were compared with a model solution of the same amino acid subjected to heating [28]. At the same time, a model solution of glutamine, the glutamic acid amide, was subjected to the same trail.

2. Materials and Methods

2.1. Grape Musts Sampling

Local wineries supplied grape musts of Lambrusco, Trebbiano toscano (from here on Trebbiano), and Spergola, some of the raw products currently used for cooked must and Balsamic vinegars production. Musts were stored at – 20 °C and defrosted immediately before use.

2.2. Grape Must Concentration

An aliquot (1 L) of each must was concentrated to half of the initial volume in partially capped 1-L conical flasks. The masses were rapidly heated to boiling point and the process was carried out by a hot plate-magnetic stirrer (MR 3003, Heidolph Elektro, Kelheim, Germany) equipped with a thermometer probe. Then the temperature was set at 90 °C for the whole process (30 h). Every 3 h, 1.5 mL of must were withdrawn and immediately stored at – 20 °C for analyses. Each concentration process and analyses were carried out in duplicate.

2.3. Model Solutions

Four model solutions were set by dissolving two different amounts of glutamic acid or glutamine into 20 mL of water and the pH was adjusted to 3.0 with diluted H_2SO_4. A set of solutions (glutamic acid 0.110 g, 0.75 mmol; glutamine 0.104 g, 0.71 mmol) was heated in closed test tubes and kept at 90 °C for 30 h. The other two solutions (glutamic acid 0.099 g, 0.68 mmol;

glutamine 0.109 g, 0.74 mmol) were heated under the same conditions, but for 3 h only. Samples (1.5 mL each) were collected at 0 h, 18 h, and 30 h for the former two solutions, and each hour for the latter two solutions.

A pyroglutamic acid solution (pH = 3) was heated into a water bath (90 °C for 10 h) and a sand bath (270 °C for 5 min).

2.4. Reagents

Fluka Sigma–Aldrich® (Milan, Italy) supplied pure reference compounds (ammonium chloride, L-glutamic acid, L-glutamine, and DL-pyroglutamic acid) hydrochloric acid, sulfuric acid, and diethylethoxymethylenemalonate (DEEMM), while VWR International S.r.l. (Milan, Italy) supplied high-purity solvents (acetonitrile and methanol). A Milli-Q purification system (Millipore, Milan, Italy) provided deionized water.

2.5. Physico-Chemical Determinations

Brix were measured in accordance with EU community methods for the analysis of wines [29].

2.6. Determinations of Amino Acids and Other Nitrogen Substances

Peak identification was carried out by comparing retention times of pure standards. Quantification was performed through an external standard calibration method.

2.6.1. HPLC Determination of Amino Acids and Other Nitrogen Substances

A pre-column derivatization method with DEEMM was used for the quantification of amino acids and ammonium ion [30] in their aminoenone form. Aminoenone derivatives were obtained by reaction of 30 μl of DEEMM and 0.5 ml of target sample without any pre-treatment in 0.875 ml of borate buffer 1 M (pH = 9) and 375 μl of methanol. The reaction took place into a screw-cap test tube over 30 min in an ultrasound bath. The sample was then heated at 70 °C for 2 h.

Quantification was performed with a Perkin Elmer HPLC system (Series 200 LCP) equipped with UV/Vis Detector (Perkin-Elmer series LC-295) set a 280 nm (λ_{max}). Samples were injected with a 20-µl loop using an injection valve (Rheodyne Inc., Cotati, CA) onto a Novapak C_{18} (30 cm × 3.9 mm i.d.; 4 µm p.s.), with a flow rate of 0.9 ml/min using a linear gradient elution [30].

2.6.2. HPLC Determination of Pyroglutamic Acid

As DEEMM is unable to react with amidic nitrogen, pyroglutamic acid was separated by anion-exclusion HPLC modifying a method used for organic acid determination [31, 24]. A Bio-Rad Aminex HPX-87H hydrogen-form cation exchange resin-based column (300 × 7.8-mm i.d.) operated at 50 °C, with a flow rate of 0.5 mL/min using a solvent system (H_2SO_4 0.045 N adjusted to pH 1.35; CH_3CN 10 %). UV detection (200 nm) was used for pyroglutamic acid quantification [32].

2.7. Statistical Analysis

Statistical analysis was carried out by means of the Statistica version 8.0 software (Stat Soft, Inc., Tulsa, U.S.A.).

3. Results and Discussion

3.1. Cooking Model System

The optimization of the cooking process was carried out using a 20%-sucrose solution that was concentrated in a conical flask partially capped with aluminum foil in order to prolong process as in real conditions. Other laboratory vessels (crystallizer and a wide open conical flask) were tested, but they reached the desired concentration too quickly, and for this reason were rejected.

In order to limit the perturbation of the system, the sample withdrawal was as low as possible. A total amount of only 16.5 ml of each must was consumed at the end of the cooking time.

3.2. Calibration and Method Optimization

Equations for external standard calibration were calculated by peak areas of standard solutions. Linearity in the concentration range was satisfactory ($R^2 > 0.996$ for all the substances). The detection limits were 0.55 ppm for pyroglutamic acid, 0.13 ppm for glutamine, 0.15 ppm for glutamic acid, and 0.05 ppm for NH_4^+.

For pyroglutamic acid, retention time optimization was carried out by modifying acetonitrile content and pH of the mobile phase. Acetonitrile shortened the retention time of pyroglutamic acid run, passing from 26 min (no CH_3CN) to 22 min with 10% CH_3CN. Working close to the highest operative limit of the column (pH = 3), a further 2-min reduction of retention time was observed.

3.3. Degradation of Glutamic Acid in Heat-Concentrated Grape Musts

Increase of Glu content, as a consequence of water evaporation, and decrease of Glu concentration for its thermal degradation, influenced the final Glu amount.

The study of Glu decomposition was carried out using two methodologies: (i) the ratio between the concentration of Glu and the corresponding °Brix at the beginning (t_0) and at the end of the process (t_{30}), and (ii) the autoscaled values of Glu concentrations divided by the corresponding °Brix plotted against time.

3.3.1. Concentration Ratio

As Glu strongly degrades during the process, while sugars are considerably more stable, the rate of concentration of Glu during the process compared to a reference concentration parameter (°Brix) has been expressed as concentration ratio [C ratio]:

$$[\text{C ratio}] = (\text{Glu}_{t0}/°\text{Brix}_{t0}) \,/\, (\text{Glu}_{t30}/°\text{Brix}_{t30})$$

that is

$$[\text{C ratio}] = (\text{Glu}_{t0}/\text{Glu}_{t30}) \,/\, (°\text{Brix}_{t30}/°\text{Brix}_{t0}) \qquad (1)$$

where Glu_{t0}, Glu_{t30}, $°Brix_{t0}$, and $°Brix_{t30}$ are the Glu and °Brix concentration mean values at the beginning (t_0) and at the end (t_{30}) of the cooking process, respectively. This data processing eliminated the effect of water evaporation.

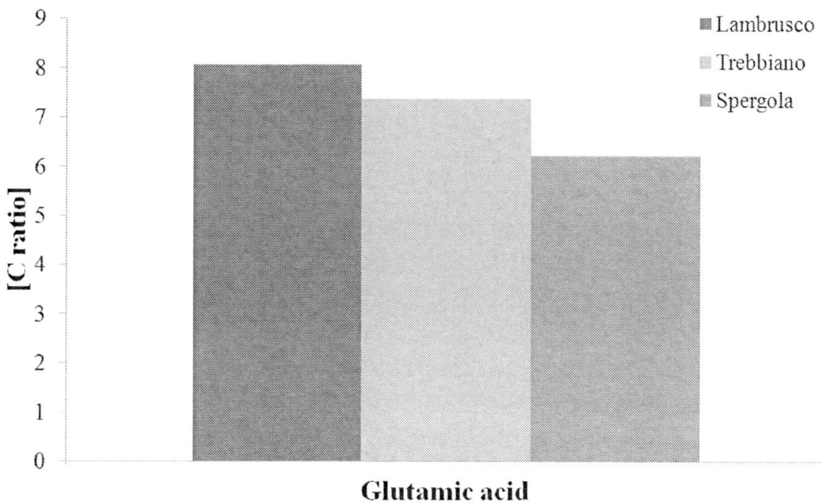

Figure 2. Concentration ratio [C ratio] for glutamic acid.

A result of Eq. (1) close to 1 indicates a negligible Glu degradation, while values > 1 indicates Glu degradation that is as higher as the value is far from 1. Finally, values < 1 would show that amino compound accumulation overtakes the normal course of must concentration.

Glutamic acid showed figures quite high (> 6), but not constant among the musts (Figure 2). In order to bypass the limits of this approach, focused only on the initial and the final stage, a second approach was carried out.

3.3.2. Comparison of Slopes

The slopes of the straight lines obtained by linear regression of the autoscaled data give information about the fate of the amino acid during the whole cooking process. A positive sign of the slope implies the predominance of amino acid concentration effect, while degradation is indicated by a negative sign. A value of the slope near to zero reveals a balance between concentration and degradation, i.e. a constant concentration. In order to eliminate the presence of false difference as a consequence of different amino acid concentrations, autoscaled values $[A_{aut}]$ were used. Data autoscaling was carried out using the following expression:

$$[A_{aut}] = (aa - aa_{mean})/aa_{sd}$$

where aa is the ratio between the Glu concentration and the corresponding °Brix, aa_{mean} is the mean of Glu during the whole process for each must, and aa_{sd} is its standard deviation.

Data autoscaling reduces the real values into a comparable set, thus eliminating the effect of the higher initial content, but keeping unchanged data trend.

The case of glutamic acid is very explicative when its concentration is plotted using raw (amino acid concentration/°Brix vs. time; Figure 3A) or autoscaled values (Figure 3B). The trends fit with straight lines for all the musts and show as Glu decreased in a regular way, tending to zero at time 30 h. The coefficient of determination (R^2) is always > 0.95.

Due to the fact that pyroglutamic acid can derive from different routes (Figure 1), it is not possible to determine pyroglutamic acid obtained by Glu conversion in sample containing also glutamine.

3.4. Degradation in Model Solutions

To gain a deeper comprehension of Glu degradation and conversion, similar experiments were carried out with model solutions, where temperature and pH were perfectly comparable to experimental must cooking conditions. However, musts in real concerns are cooked in open vessels to allow their concentration, while these experiments were carried out in closed tubes to highlight the reactivity of the system, only. To emulate the real process, the time was the same of the first set of samples (30 h). At the same way, to gain a better interpretation of the first hours of the process, a second set of model solutions was set up. In this case, the process was followed for 3 h only, with a closer sampling withdrawal.

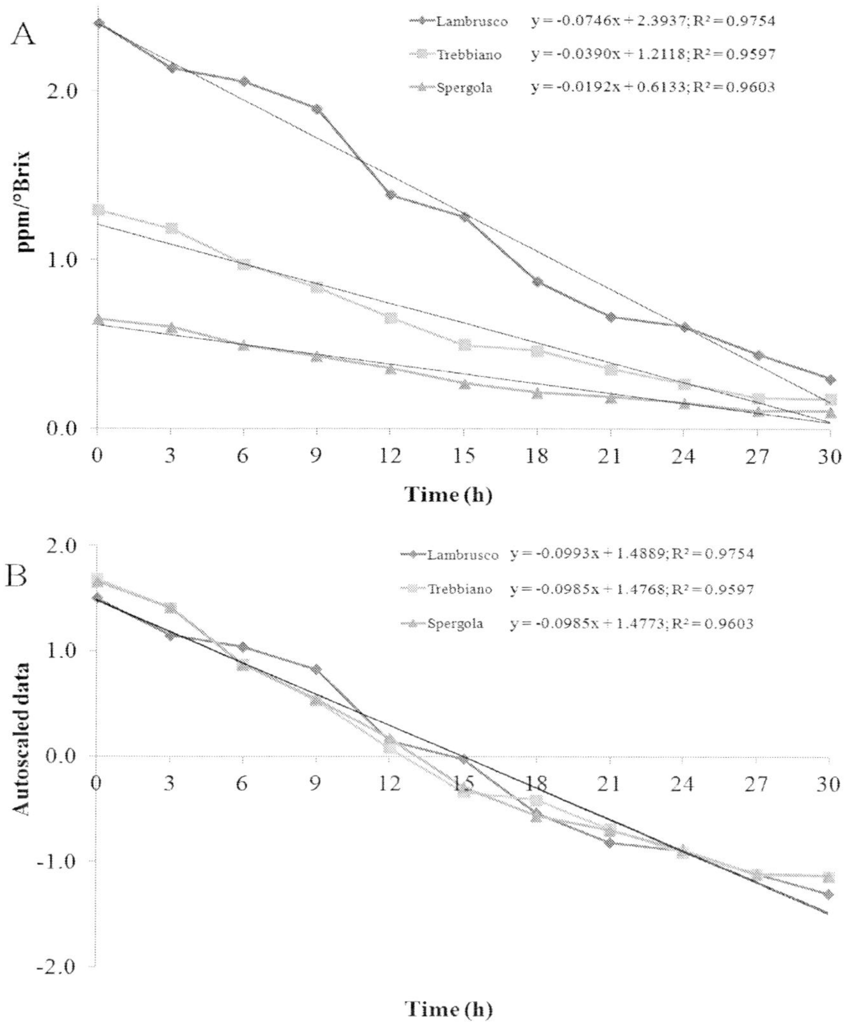

Figure 3. Regression straight lines for glutamic acid concentration vs. time of cooking in the three different musts: Lambrusco, Trebbiano, and Spergola. In A, raw data were plotted vs. time, while in B, autoscaled data were used.

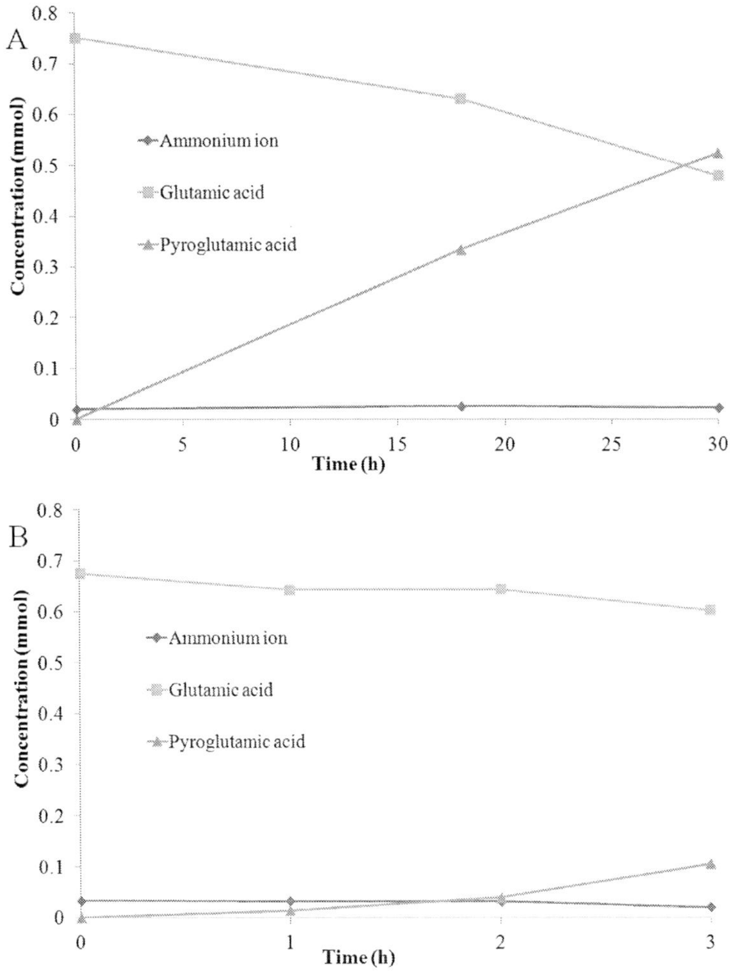

Figure 4. Glutamic acid degradation in model solution. 30-h experiment (A) and 3-h experiment (B).

On the contrary, glutamine (Figure 5A) disappeared with an almost quantitatively transformation into pyroglutamic acid and ammonia. Finally in model solutions, the amino acid concentrations were deliberately higher than those ones detected in the musts to have a more marked vision of the phenomena. Glu slowly degraded into pyroglutamic acid with a negligible amount of ammonia production (Figure 4A and B). The same behavior was shown in 3-hour experiment (Figure 5B), confirming a quick and regular glutamine loss. Finally, a slight tendency to glutamine hydration to yield glutamic acid (Figure 5A inset and 5B) was evident in both cases, as well as already observed by Archibald [17]. Pyroglutamic acid showed the characteristic of a final product. In fact, a further thermal stress of this substance had negligible consequences. Pyroglutamic acid heated for 10 h at 90 °C gave no concentration modifications, while a drastic treatment (270 °C for 5 min) caused a 16% loss. However, a reaction with other constituents may not be excluded in real samples such as musts, vinegars, or wines. For example, Webb et al. [33] reported the presence of pyroglutamic ethyl ester (ethyl 5-oxopyrrolidine-2-carboxylate) in Sherry wines, while in vinegars the presence of N-acetyl-pyroglutamic acid (1-acetyl-5-oxopyrrolidine-2-carboxylic acid) could be likely.

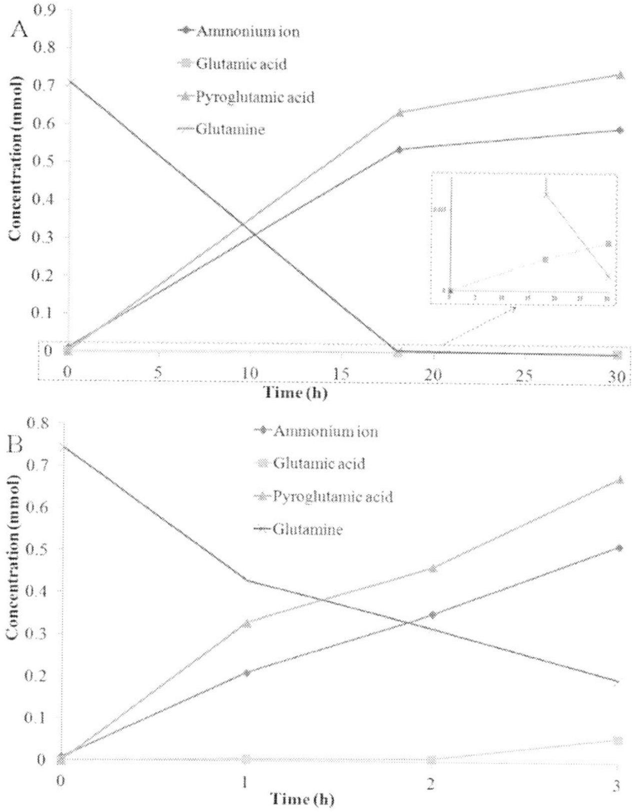

Figure 5. Glutamine degradation in model solution. 30-h experiment (A) and 3-h experiment (B). In A, inset is an expansion of the 0-0.007 mmol range that highlights glutamic acid and glutamine trends.

Conclusion

In conclusion, glutamic acid, one of the most representative amino acids in food, is the main responsible of umami, that is perceived by human taste receptors as an indicator of amino acid presence in food.

When an acidic solution of glutamic acid is heated, this amino acid progressively degraded throughout all the process. This trend was confirmed by a model solution, that also showed as pyroglutamic acid was the main product of degradation.

References

[1] Ritthausen, H. (1886). Ueber die Glutaminsäure. *Journal für Praktische Chemie*, 99(1), 454–462.

[2] Wolff, L. (1890). Ueber glyoxylpropionsäure und einige abkömmlinge derselben. *Justus Liebigs Annalen der Chemie*, 260(1), 79–136.

[3] The Merck Index (1996). Glutamic acid, 760–761. In: S. Budavari (ed.) The Merck Index. An Encyclopedia of Chemicals, Drugs, and Biologicals (12th edition). Whitehouse Station, N. J., U.S.A.: Merck and Co., Inc.

[4] Ikeda, K. (1908). Japanese patent 14805.

[5] Ikeda, K. (1909). New Seasonings. *Journal of the Chemical Society of Tokyo*, 30, 820–836.

[6] Yamaguchi, S., and Ninomiya, K. (2000). Umami and food palatability. *Journal of Nutrition*, 130S, 921–926.

[7] Bufe, B., and Meyerhof, W. (2006). The human perception of taste compounds (Chapter 1). In: A. Voilley and P. Etievant (eds.), *Flavour in food*. Boca Raton, FL, U.S.A: CRC Press.

[8] Sugita Y. (2002). Flavor Enhancers (Chapter 14). In: A. L. Branen, P. M. Davidson, S. Salminen, J. H. Thorngate III (eds.) *Food Additives* (2nd edition). New York, U.S.A.: Marcel Dekker, Inc.

[9] Kawakita, T. (2000). L-Monosodium glutamate (MSG). In: Kirk-Othmer Encyclopedia of Chemical Technology, Vol. 2 (4th edition, 410-421). New York, U.S.A.: Wiley.

[10] Hegenbart, S. L. (1998). Alternative enhancers. *Food Product Design*, 7(11), 60–71.

[11] Nagodawithana, T. (1992). Yeast-derived flavors and flavor enhancers and their probable mode of action. *Food Technology*, 11, 138–144.

[12] Chandrashekar, J., Hoon, M. A., Ryba, N. J. P., and Zuker, C. S. (2006). The receptors and cells for mammalian taste. *Nature*, 444, 288–294.

[13] Fuke, S., and Ueda, Y. (1996). Interactions between umami and other flavor characteristics. *Trends in Food Science and Technology*, 71, 407–411.

[14] Jinap, S., and Hajeb, P. (2010). Glutamate. Its applications in food and contribution to health. *Appetite*, 55(1), 1–10.

[15] Raghavan, S. (2007). Trends in the World of Spices Today (Chapter 2). In: Handbook of Spices, Seasonings, and Flavorings (2nd edition). Boca Raton, FL, U.S.A.: CRC Press.

[16] Nunes, R. S., and Cavalheiro, É. T. G. (2007). Thermal behavior of glutamic acid and its sodium, lithium and ammonium salts. *Journal of Thermal Analysis and Calorimetry*, 87(3), 627–630.

[17] Archibald, R. M. (1945). Chemical characteristics and physiological rôles of glutamine. *Chemical reviews*, 37(2), 161–208.

[18] Vickery, H. B., Pucher, G. W., Clark, H. E., Chibnall, A. C., and Westall, R. G. (1935). The determination of glutamine in the presence of asparagine. *Biochemical Journal*, 29(12), 2710–2720.

[19] Airaudo, C. B., Gayte-Sorbier, A., and Armand, P. (1987). Stability of glutamine and pyroglutamic acid under model system conditions: Influence of physical and technological factors. *Journal of Food Science*, 52(6), 1750–1752.

[20] Sohn, M. and Ho, C. T. (1995). Ammonia generation during thermal degradation of amino acids. *Journal of Agricultural and Food Chemistry*, 43(12), 3001–3003.

[21] Gayte-Sorbier, A., Airaudo, C. B., and Armand, P. (1985). Stability of glutamic acid and monosodium glutamate under model system conditions: influence of physical and technological factors. *Journal of Food Science*, 50(2), 350–352.

[22] Belitz, H. D., Grosch, W., Schieberle, P. (2009). Amino acids, peptides, proteins (Chapter 1). In: *Food Chemistry* (4th ed.). Berlin, Heidelberg, Germany: Springer-Verlag.

[23] Medicamenta (1996). VII edizione. Acido L-pirrolidoncarbossilico. Milano, Italy: Cooperativa Farmaceutica.

[24] Montevecchi, G., Masino, F., Chinnici, F., and Antonelli, A. (2010). Occurrence and evolution of amino acids during grape must cooking. *Food Chemistry*, 121, 69–77.

[25] Rivero-Pérez, M. D., Pérez-Magariño, S., and González-San José, M. L. (2002). Role of melanoidins in sweet wines. *Analytica Chimica Acta*, 458(1), 169–175.
[26] Piva, A., Di Mattia, C., Neri, L., Dimitri, G., Chiarini, M., and Sacchetti, G. (2008). Heat-induced chemical, physical and functional changes during grape must cooking. *Food Chemistry*, 106(3), 1057–1065.
[27] Repubblica Italiana (2007). Decreto Direttoriale 19/06/2007. Settima revisione dell'elenco nazionale dei prodotti agroalimentari tradizionali. *Gazzetta Ufficiale della Repubblica Italiana*, 147 (27 June) (S.O. n. 146).
[28] Montevecchi, G., Masino, F., and Antonelli, A. (2011). Pyroglutamic acid development during grape must cooking. *European Food Research and Technology*, 232, 375–379.
[29] EEC (1990). Commission Regulation 2676/1990. Community methods for the analysis of wines. *Official Journal of the European Community,* L272, 1–192.
[30] Gómez-Alonso, S., Hermosín-Gutiérrez, I., and García-Romero, E. (2007). Simultaneous HPLC analysis of biogenic amines, amino acids, and ammonium ion as aminoenone derivatives in wine and beer samples. *Journal of Agricultural and Food Chemistry*, 55(3), 608–613.
[31] Castellari, M., Versari, A., Spinabelli, U., Galassi, S., and Amati, A. (2000). An improved HPLC method for the analysis of organic acids, carbohydrates, and alcohols in grape musts and wines. *Journal of Liquid Chromatography and Related Technologies*, 23(13), 2047–2056.
[32] Shih, F. F. (1985). Analysis of glutamine, glutamic acid and pyroglutamate in protein hydrolysates by HPLC. *Journal of Chromatography A*, 322, 248–256.
[33] Webb, A. D., Kepner, R. E., and Maggiora, L. (1967). Sherry aroma. VI. Some volatile components of flor sherry of Spanish origin. Neutral substances. *American Journal of Enology and Viticulture*, 18(4), 190–199.
[34] Resources Council, Science and Technology Agency, Japan (1986). Standard tables of food composition in Japan. Amino acid composition of foods.
[35] The Glutamate Association. http://www.msgfacts.com/
[36] Coultate, T. P. (2002). Proteins (Chapter 5). In: Food. The chemistry and its components (4th edition). Cambridge, U.K.: The Royal Society of Chemistry.

[37] U.S. Department of Agriculture (1984). Composition of Foods: Vegetables and Vegetable Products. *Agriculture Handbook*, 8-11.

[38] Rassin, D. R., Sturman, J. A., and Gaull, G. E. (1978). Taurine and other free amino acids in milk of man and other mammals. *Early human development*, 2(1), 1–13.

[39] Maeda, S., Eguchi, S., and Sasaki, H. (1961). The content of free L-glutamic acid in various foods (Part 2). *Journal of Home Economics* (Japan), 12, 105.

[40] Maeda, S., Eguchi, S., and Sasaki, H. (1958). The content of free L-glutamic acid in various foods. *Journal of Home Economics,* (Japan), 9, 163–167.

[41] Kiuchi, T., and Kondo, Y. (1984). The study of free amino acids and related compounds in vegetable foods. *Report of Hiroshima Women's University*, 20, 65.

In: Glutamic Acid
Editors: D. Balcazar and E. Perez
ISBN: 978-1-62257-236-6

Chapter 5

Application of a Natural Biopolymer Poly (γ-Glutamic Acid) as a Bioflocculant and Adsorbent for Cationic Dyes and Chemical Mutagens: An Overview*

***B. Stephen Inbaraj*† *and B. H. Chen*≠**
Department of Food Science, Fu Jen University, Taipei, Taiwan

Abstract

Poly (γ-glutamic acid) (γ-PGA), a novel polyanionic and multifunctional macromolecule synthesized by Bacillus species, has attracted considerable attention because of its eco-friendly, biodegradable and biocompatible characteristics. Recently, its application in a wide range of fields such as food, agriculture, medicine, hygiene, cosmetics and environment has been explored. This book chapter reviews the

---

* A version of this chapter also appears in *Biochemical Engineering*, edited by Fabian E. Dumont and Jack A. Sacco, published by Nova Science Publishers, Inc. It was submitted for appropriate modifications in an effort to encourage wider dissemination of research.

† E-mail address: sinbaraj@yahoo.com.
≠ E-mail address: 002622@mail.fju.edu.tw. Tel.: 886-2-29053626, Fax: 886-2-29053415.

literature reports on the application of γ-PGA as a flocculating agent, and adsorbent for cationic dyes and chemical mutagens, affected by several process parameters including pH, temperature, contact time, metal cations, concentration and molecular weight of γ-PGA.

1. Introduction

Biopolymers are molecules produced by biological cells including bacteria, fungi, plant and animal cells. Poly (γ-glutamic acid) (γ-PGA) is a natural and water-soluble biopolymer produced by a variety of *Bacillus* species through fermentation. The γ-PGA was first discovered as a major constituent in the capsule of *Bacillus anthracis*, which was subsequently released into the growth medium upon autoclaving or aging and autolysis of the cells [1, 2]. In a later study, Fujji [3-5] reported fermented soybean (Natto), a traditional health food in Japan, also contained γ-PGA in the form of viscous sticky mucilage, with *B. subtilis* natto being responsible for its synthesis. Ever since Bovarnick [6] reported the free secretion of γ-PGA in the growth medium of *B. subtilis*, several researchers have explored the extracellular synthesis of γ-PGA from a variety of *Bacillus* species [5]. Although most studies on γ-PGA was carried out between 1950s and 1970s [7-13], an increasing attention was drawn in recent years mainly because of its biodegradable and biocompatible properties. The γ-PGA is primarily composed of repetitive glutamic acid monomer units, which are connected by γ-amide linkages between α-amino and γ-carboxyl groups and are thus synthesized in a ribosome-independent manner [5, 14, 15] (Figure 1). The naturally-produced γ-PGA contains nearly equal proportion of D- and L-glutamic acid units, yet the ratio of two optical isomers can be partially controlled by technological methods, yielding γ-PGA with different degree of stereoselectivity [5, 14-16]. In addition, γ-PGA can be synthesized in different salt forms (Na, K, Ca, Mg and NH_4) and varying molecular weights (10,000 to 2 million Daltons) [5, 14, 15]. Owing to its non-toxic, polyanionic and multifunctional characteristics, γ-PGA finds potential application in a wide range of fields such as food, agriculture, cosmetics, medicine and environment [5, 14, 15]. Some specific applications of γ-PGA include its use as a health food, thickener, humectant, bitterness-relieving agent, osteoporosis-preventing agent, cryoprotectant, drug carrier, sustained release material, curable biological adhesive, biodegradable fibers, hydrogel (super water absorbent), moisturizer, tissue engineering material, biodegradable packing material, dispersant, flocculant, adsorbent of toxic cations and chemical mutagens, animal feed additive, enzyme immobilizing material, liquid display and conductive display materials [17-32]. This book chapter intends to review the reported studies on the application of γ-PGA as a bioflocculant, and an adsorbent for scavenging cationic dyes and chemical mutagens.

2. Biosynthesis and Physico-chemical Properties of γ-PGA

2.1. Biosynthesis

Several *Bacillus* species produce γ-PGA as a capsular component or an extracellular viscous material outside the cell body for subsequent release into the fermentation broth. To enhance the productivity of γ-PGA, researchers have investigated the nutrient requirements [5], which varied according to the type of *Bacillus* strain. Based on the nutrient requirements, γ-PGA-producing bacteria are divided into two groups: one requires the addition of L-glutamic acid (*de nova* method) to the medium to stimulate both cell growth and γ-PGA synthesis, while the other does not (salvage bioconversion method) [5, 33-35]. Table 1 summarizes the nutrient requirements, cultivation condition, production yield, D/L glutamic acid ratio and molecular weight of γ-PGA synthesized from various *Bacillus* strains [7-9, 36-40]. Besides carbon and nitrogen sources, factors such as ionic strength, aeration and medium pH, all of which can affect the productivity, molecular weight, stereochemical composition and quality of γ-PGA [5, 41]. For instance, the molecular weight of γ-PGA synthesized from *B. licheniformis* ATCC 9945A increased by a factor of approximately 1.8 (1.2 to 2.2 million g/mol) for a concentration rise of NaCl from 0 to 4% [5, 42]. The schematic pathway proposed for synthesis of γ-PGA from *B. subtilis* IFO 3335 is depicted in Figure 2 [42]. A more detailed description on the biosynthesis of γ-PGA has been well reviewed [5, 8, 14, 15, 41, 43], and an elaborate discussion here is beyond the scope of this chapter.

2.2. Physico-chemical Properties

The γ-PGA biosynthesized from *B. subtilis* (natto) was characterized for various physical and chemical properties by using several instrumental techniques including Fourier transform infrared spectrophotometry (FT-IR), ^{1}H- and ^{13}C-nuclear magnetic resonance spectroscopy (^{1}H- and ^{13}C-NMR), differential scanning calorimetry (DSC) and thermal gravimetric analysis (TGA), and the results are furnished in Table 2 [15]. Elemental analysis of purified H-form of γ-PGA at 1.23 x 10^6 kDa (C: 44.86%; H: 5.91%; N: 10.49%; S: 0%) was in close agreement with the calculated values (C: 46.51%; H: 5.43%; N: 10.85%; S: 0%) based on the formula composition [15]. The viscosity of γ-PGA was reported to be strongly dependent on pH, γ-PGA concentration, temperature and ionic strength [15, 44], which should be due to the abrupt changes caused by these parameters on the conformation of γ-PGA. The viscosity of 4% Na-γ-PGA solution at 25°C increased following a rise in pH and γ-PGA dose, but decreased with temperature and ionic strength [15]. The γ-PGA remained unaltered on heating at 80°C for 60 min, however, it was rapidly hydrolyzed at 120°C because of random chain scission [33].

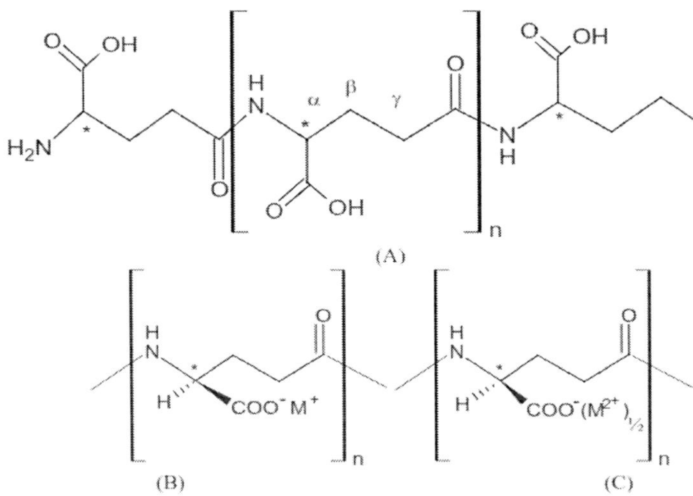

Figure 1. The chemical structure of poly (γ-D,L-glutamic acid) and poly (γ-glutamates). (A) H-form of γ-PGA; (B) monovalent metal salt form of γ-PGA; (C) divalent metal salt form of γ-PGA; $M^+ = Na^+$, K^+ or NH_4^+; $M^{2+} = Ca^{2+}$ or Mg^{2+}; Greek symbols (α, β, γ) denote carbon positions; Asterisk (*) symbol indicate the chiral carbon atom.

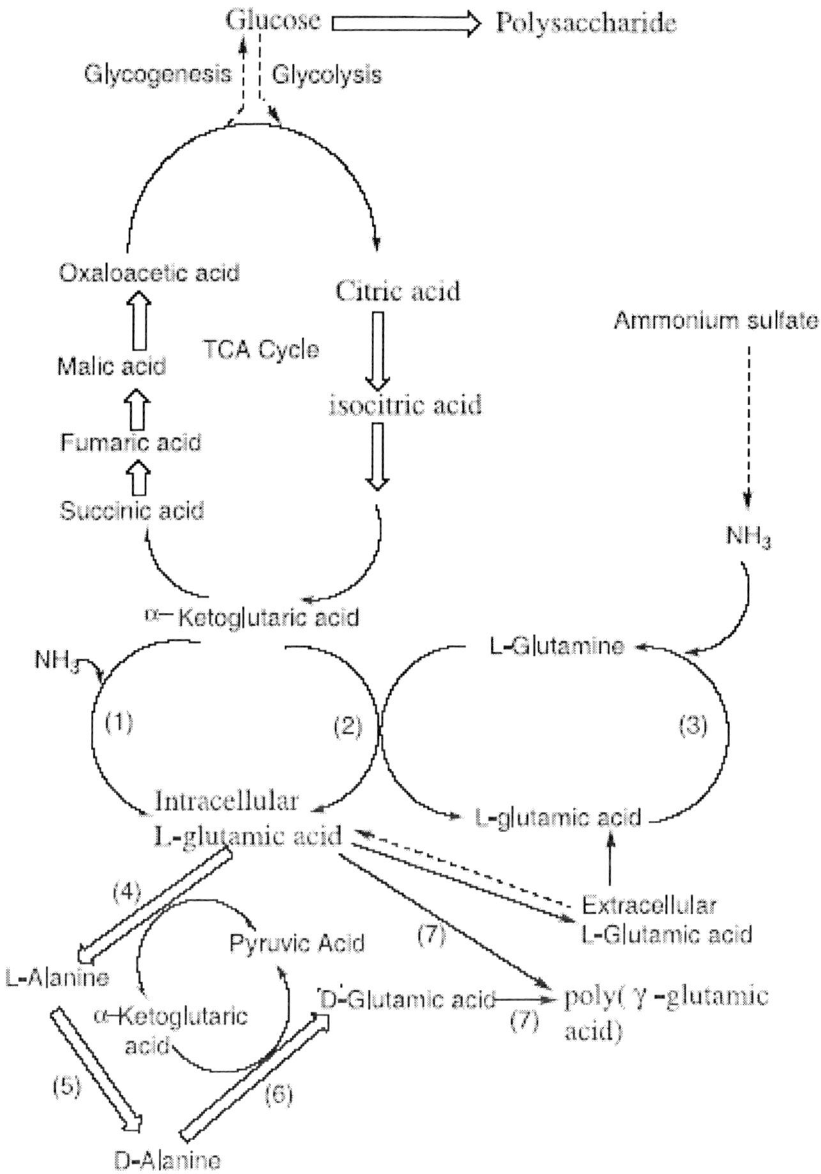

Figure 2. Proposed pathway of biosynthesis of poly (γ-glutamic acid) from *B. subtilis* IFO 3335. (1) Glutamate dehydronase; (2) Glutamate 2-oxoglutarate (α-ketoglutarate) aminotrasferase; (3) Glutamine synthetase; (4) Pyruvic acid aminotransferase (L-Glutamic acid); (5) Alanine racemase; (6) Pyruvic acid aminotransferase (D-Glutamic acid); (7) PGA synthetase (Source: Kunioka 1997).

Table 1. Synthesis of γ-PGA from various *Bacillus* strains

Bacillus strains	Nutrients	Cultivation conditions	Yield (g/L)	Molecular weight	D : L ratio	References
B. licheniformis ATCC 9945	Glutamic acid, Glycerol,	37°C, 4 days	17-23	1.4 x 104 – 9.8 x 105	45-85 : 55-15	[8]
B. licheniformis ATCC 9945	Citric acid, NH4Cl,	37°C, 4 days	17-23	1.4 x 104 – 9.8 x 105	45-85 : 55-15	[36]
B. subtilis IFO 3335	Glutamic acid (30 g/L), citric acid (20 g/L)	37°C, 2 days	10-20	1.0 x 105 – 2.0 x 106	17 : 83	[37]
B. subtilis TAM-4	Fructose (75 g/L), NH4Cl (18 g/L)	30°C, 4 days	20	6.0 x 105 – 1.6 x 106	78 : 22	[38]
B. licheniformis A35	Glucose (75 g/L), NH4Cl (18 g/L)	30°C, 3-5 days	8-12	3.0 – 5.0 x 105	59 : 41	[39]
B. subtilis F02-1	Glutamic acid (70 g/L), glucose (1 g/L), veal infusion broth (20 g/L)	30°C, 2-3 days	50	1.20 x 106	55 : 45	[34,35]
B. subtilis NRRL B2612	Wheat gluten (200 g/L)	33°C, 2-3 days	10-14	2.0 x 104	52 : 48	[7]
B. subtilis var. polyglutamicum	Glucose (50 g/L), urea (7.5 g/L)	30°C, 3-4 days	17-19	1.1 x 106	40 : 60	[9]
B. subtilis (natto) MR 141	Maltose (60 g/L), soy sauce (70 g/L), sodium glutamate (30 g/L)	40°C, 3-4 days	35	-	-	[40]
B. subtilis (natto)	Glucose (8%), sodium L-glutamate (10%)	37°C, 2 days	35	1.5 – 3.0 x 106	52 : 48	[15]

2.3. Structural Characteristics

The monomer glutamic acid in γ-PGA possesses three chemically-active functional groups in the following order of reactivity: α-NH_2 > α-COOH > γ-COOH. In a chemical-catalyzed polymerization of glutamic acid, the condensation occurs between the α-COOH and α-NH_2 groups, resulting in the formation of poly (α-glutamic acid) through α-peptide linkages. However, in the submerged fermentation process, a portion of L-glutamic acid is enzymatically racemized to D-glutamic acid and eventually both D- and L- glutamic acids are co-polymerized through a novel γ-peptide bonding between a less reactive γ-COOH and α-NH_2 groups to form poly (γ-D,L-glutamic acid). An α-peptide bond, which is normally found in protein structures, can be easily decomposed by most of the protease enzymes, whereas γ-peptide bond can be hydrolyzed only by a rarely available γ-glutamyltranspeptidase and γ-PGA is thus considerably resistant to microbial attack [5, 14, 15]. Likewise, poly (γ-glutamic acid) contains four intramolecular hydrogen bonds (3_{19}, 3_{17}, 3_{14} and 3_{12}), formed between the carbonyl group of one γ-peptide bond and the amino group of another γ-peptide bond within every 3 glutamic acid moieties [15, 45], as opposed to only one hydrogen bond within an average of 3.6 amino acid residues in proteins found in nature. Optical rotatory dispersion (ORD) studies [15] have shown these strong hydrogen bonds in γ-PGA to be responsible for the compact α-helix conformation rendering a strong hydrophobic character (insoluble nature) at pH 2.0. But, as pH rises, the hydrogen bonding breaks down and the α-helix conformation is converted into a linear random-coil conformation resulting in ionization of α-COOH groups [5, 14, 15]. At pH = pK_a (4.09, determined by potentiometric titration), about 50% of α-COOH groups are ionized to α-COO^- anions (50% polyanionic random-coil transition). Beyond pH 6.0, γ-PGA mainly exists in the linear random-coil conformation and exhibits polyanionic characteristics, making it a novel multi-functional biopolymer for application in a wide range of fields.

3. γ-PGA as a Bioflocculant

Flocculation is a process of promoting the agglomeration of smaller particles into larger and more easily sedimentable flocs. Flocculating agents have been frequently used in wastewater treatment, food and fermentation industries, drinking water treatment and industrial downstream processing [46-50]. Among various flocculants, synthetic organic flocculants (polyacrylamide derivatives, polyacrylic acids and polyethylene imine) are widely applied due to their low cost and high efficiency [51]. Nevertheless, their use often gives rise to environmental and health problems as they are not readily biodegradable and some of their monomers such as acrylamide are neurotoxic and potential carcinogens [52-54]. Thus, the development of safe, biodegradable flocculants is imperative to reduce the health risks associated with synthetic flocculants. This chapter subsection unveils the literature reports on flocculation capability of γ-PGA as affected by pH, temperature, γ-

PGA concentration, metal cations, cross-linking γ-PGA and molecular weight of γ-PGA.

3.1. Assay of Flocculation Activity and Flocculation Rate

Measurement of flocculation activity was based on decrease in turbidity through addition of γ-PGA in a reaction mixture containing test suspension and metal cations [30, 31, 54, 55]. After the visible flocs were settled, the optical density (OD) of the supernatant of reaction mixture with or without γ-PGA (control) was measured at 550 or 660 nm using a spectrophotometer, and the flocculation activity and flocculation rate were calculated using the following equations:

$$\text{Flocculation activity}\,(1/\text{OD}) = \frac{1}{\text{OD}_{\text{sample}}} - \frac{1}{\text{OD}_{\text{control}}} \quad (1)$$

$$\text{Flocculation rate}\,(\%) = \frac{\text{OD}_{\text{control}} - \text{OD}_{\text{sample}}}{\text{OD}_{\text{control}}} \text{ x } 100 \quad (2)$$

3.2. Effect of γ-PGA Concentration and Molecular Weight

When evaluated at different concentrations (5-100 mg/L), the γ-PGA from PY-90 and IFO3335 strains of *B. subtilis* showed an optimum flocculation activity of 15 and 10.5 l/OD respectively, for a 20-mg/L kaolin suspension containing 4.5 and 10 mM $CaCl_2$ [54, 55]. However, Shih et al. [30] found the optimum flocculation (8.5 l/OD) to occur at a γ-PGA dose of 3.7 mg/L in a 5 g/L kaolin suspension containing 9.0 mM $CaCl_2$. This variation in γ-PGA level may be attributed to different *Bacillus* species and culture conditions employed in the latter study [30], in which *B. licheniformis* (CCRC 12826) and a combination of glutamic acid, citric acid and glycerol was used as carbon source, as opposed to *B. subtilis* and only glutamic acid used in the former study [54, 55]. Recently, Wu et al. [31] examined the flocculation activity at various concentrations of γ-PGA from DYU1 *B. subtilis* strain (10-80 mg/L) in a reaction mixture containing kaolin and 5 x 10^{-3} mM of metal cations (Ca^{2+}, Mg^{2+}, Fe^{2+}, Fe^{3+} or Al^{3+}) and a level ranging from 10-30 mg/L was found effective in attaining the optimum flocculation activity (Figure 3).

The molecular weight (MW) of γ-PGA also showed a significant impact on the flocculation activity. A high MW γ-PGA (2×10^6 kDa) from *B. licheniformis* (CCRC 12826) exhibited a higher flocculation activity compared to the purified low MW γ-PGA ($\sim 1 \times 10^5$ kDa) from Sigma [30]. Apparently, a high MW polymer contains a large number of free functional groups, which may act as bridges connecting numerous suspended particles to form flocs of large size during the flocculation reaction [49].

3.3. Effect of Added Metal Cations and pH

Incorporation of metal cations plays a key role in promoting flocculation, probably because of decline in charge density by cations leading to inter-particle bridging between suspended particles. Flocculation studies on bacterial γ-PGA isolate were unequivocal in reporting the synergistic effect of added metal cations, with the trivalent ions (Al^{3+} or Fe^{3+}) exerting the highest effect, followed by divalent (Ca^{2+}, Mg^{2+} or Fe^{2+}) and monovalent (Na^+ or K^+) ions [30, 31, 55]. The amount of metal ions required for maximum flocculating efficiency decreased following an increase in the valency of metal ions. Wu et al. [31] reported an optimum concentration of >10 mM for monovalent ions, but only 0.10-0.90 mM for divalent ions and <0.005 mM for trivalent ions in a reaction mixture of kaolin and γ-PGA from *B. subtilis* (DYU1) (Figure 4). Nonetheless, a high Ca^{2+} concentration of 2-8 mM and 13.5 mM was required to attain the maximum flocculation activity for γ-PGA produced from *B. subtilis* (PY-90) and *B. licheniformis* (CCRC 12826), respectively [30, 54]. Unlike monovalent and divalent ions, a rise in concentration of trivalent ions (Al^{3+} or Fe^{3+}) sharply reduced the flocculation activity (Figure 4), which may be ascribed to the drop in solution pH caused by precipitation of these ions. However, the flocculation activity could be restored by adjusting pH of the reaction mixture to 7 as reported by Yokoi et al. [55], who proposed the hydroxide precipitates of trivalent ions and Fe^{2+} ions to be effective in stimulating the flocculation activity of γ-PGA at or near neutral pH. Likewise, for both Ca^{2+} and Mg^{2+} cations, the pH around 7 was shown to be optimum in achieving the largest flocculation activity and flocculating rate by γ-PGA from *B. subtilis* DYU1 (Figure 5) [30]. On the contrary, a slightly acidic pH around 4 was reported for Ca^{2+} ions to reach the maximum flocculating effect in the suspension mixture containing kaolin and γ-PGA from *B. subtilis* IFO 3335 [54, 55]. The reason for this disparity is unclear.

3.4. Effect of Temperature

The flocculation activity declined linearly with a rise in incubation temperature (30-120°C) of the reaction mixture containing kaolin suspension (5 g/L), metal cation (5 x 10^{-3} mM) and γ-PGA isolate from *B. subtilis* DYU1 (40 mg/L) [31]. The optimum flocculation activity (5.2 l/OD) and flocculation rate (90%) were attained at 30°C. But, the flocculation activity dropped by 50% after incubation at 60°C for 15 min, followed by a complete inactivation of γ-PGA at 120°C. Yokoi et al. [54] also reported a rapid decline in flocculation activity of γ-PGA from PY-90 *B. subtilis* strain at elevated temperatures, diminishing to zero on heating at 100°C (40 min). Obviously, the bioflocculants composed of protein or peptide backbone are susceptible to heat treatment, while those containing sugars and polysaccharides are thermally stable [56].

3.5. Effect of Various Inorganic and Organic Suspensions

The γ-PGA from different *Bacillus* species was not only effective in flocculating kaolin suspension, but also in organic suspensions and some other inorganic suspensions including active carbon, solid soil and acid clay (Table 3) [30, 55]. The γ-PGA isolate from *B. subtilis* (IFO3335) at 20 mg/L exhibited a greater flocculating efficiency (l/OD) in acid clay suspension (10.9-34.7) compared to that in active carbon (2.8-3.3) and solid soil (1.0-1.9). In the absence of any added metal cations, γ-PGA flocculated suspensions of calcium and magnesium salts, with the highest flocculation activity (l/OD) being shown in $Ca(OH)_2$ suspension (21.3), followed by $MgCO_3$ (9.3), $Mg(OH)_2$ (8.1), Al_2O_3 (6.1), $[Ca(PO4)_2]_3 \cdot Ca(OH)_2$ (5.7) and $CaCO_3$ (2.0) [30, 55]. The γ-PGA from *B. licheniformis* (80 mg/L) showed a good flocculating effect in organic suspensions (5 g/L) as well, which equaled to 3.0, 2.6, 4.6 and 2.8 l/OD for cellulose powder, carboxymethylcellulose, *Saccharomyces cerevisiae* and *B. circulans*, respectively (Table 3) [30].

3.6. Effect of Crosslinking Poly (γ-Glutamic Acid)

Crosslinking of γ-PGA (CL-γ-PGA) with an optimum γ-irradiation dose of 20 kGy was shown to generate the highest water absorption capacity (1005.6 mL/g) and viscosity (3.31 η) and possess twice the flocculation activity (5.48 1/OD) when compared to a non-crosslinked γ-PGA (2.74 1/OD) in kaolin suspension at 30°C and pH 5.0 [57]. However, a pretreatment with polyacrylamide (PAC) drastically raised the flocculation activity of CL-γ-PGA in different suspensions [57, 58]. Table 4 summarizes the flocculation activity of γ-PGA as affected by pretreatment with different doses of PAC for various suspensions at 30°C and pH 5.0 [58]. Compared to the control (only CL-γ-PGA), the flocculation activity (1/OD) rose by 39.9, 32.8, 30.0, 23.5, 88.2 and 12.2 for kaolin, bentonite, diatomaceous earth, *E. coli*, *M. aeruginosa* and Mandai pond (real water sample) suspensions, respectively, for PAC and CL-γ-PGA at a dose (μg/mL) of 0.5 and 10, 2 and 5, 0.5 and 10, 0.5 and 10, 0.5 and 5, and 2 and 20 [58]. Apparently, the Al^{3+} ion in PAC could electrostatically interact with the negatively charged cell surface of *E. coli* and *M. aeruginosa* or with the silicate anion induced on the surface of kaolin, bentonite and diatomaceous earth suspensions to form small flocks, which may eventually become larger following the addition of CL-γ-PGA. Yet, the pretreatment with PAC did not show any effect on the real sample suspension from Yamato River (a dirtiest river in Japan). In contrast, a reversed trend was shown in crystal violet (CV) dye suspension, probably due to electrostatic interaction between the iminium cation in CV and COO^- in CL-γ-PGA. Among various added metal cations, the trivalent ions (Al^{3+} and Fe^{3+}) demonstrated a marked effect on flocculation activity of CL-γ-PGA, whereas the divalent ions (except Mn^{2+}) and monovalent ions did not show any effect [57].

Under the optimum conditions of CL-γ-PGA and PAC, the flocculating rate expressed in "decrease in turbidity" increased linearly with incubation time for bentonite suspension, while it was rapid in the initial 5 min followed by a gradual decline for diatomaceous earth (Figure 6) [58]. This may be caused by the faster precipitating behavior of diatomaceous earth compared to bentonite. On the other hand, for *E. coli* suspension, no flocculation was observed until 5 min, but increased with time thereafter [58]. However, a sharp decline in turbidity was noticed within 5 min for *M. aeruginosa* and then reached a plateau (Figure 6).

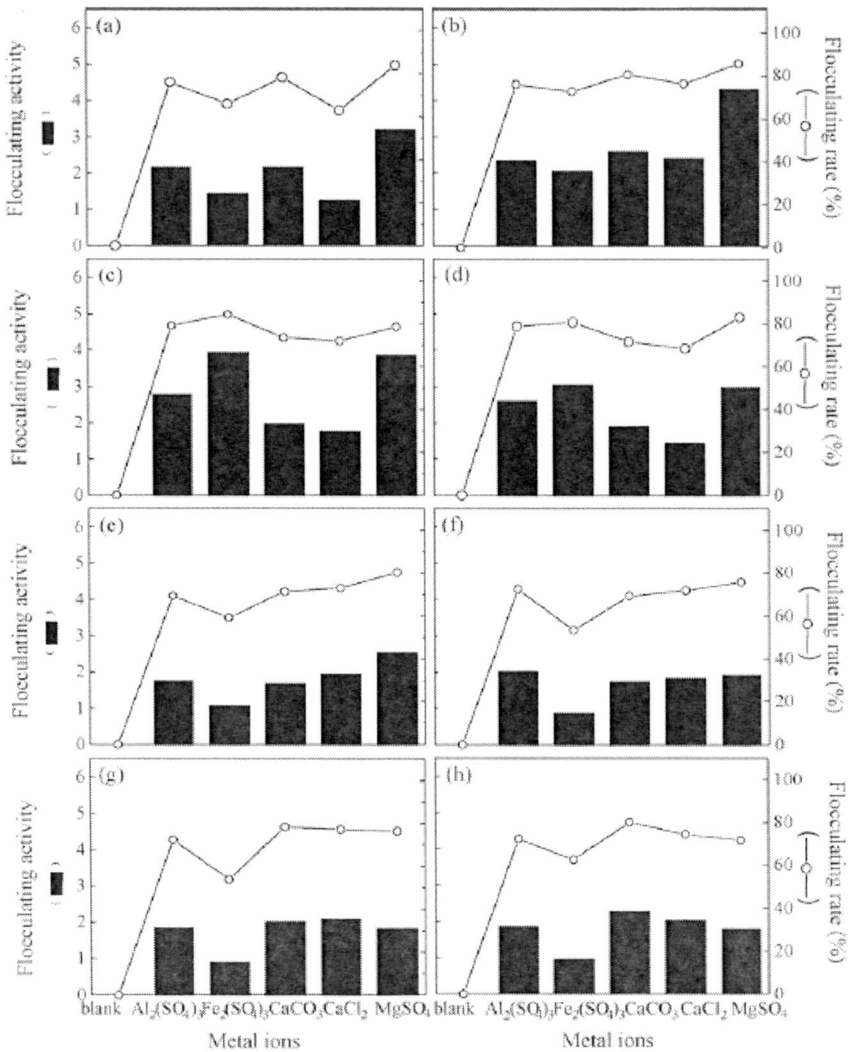

Figure 3. Effect of γ-PGA concentration on flocculation activity and flocculating rate at low metal concentration (5 x 10-3 mM) and pH 7.0. γ-PGA concentration (mg/L): (a) 10; (b) 20; (c) 30; (d) 40; (e) 50; (f) 60; (g) 70; (h) 80 (Source: Wu and Ye 2007).

Table 2. Physical and chemical parameters of γ-PGA characterized by several instrumental techniques (Source: Ho et al. 2006)

Instrumental analysis	Form of poly (γ-glutamic acid)[a]					
	H^+	Na^+	K^+	NH_4^+	Ca^{2+}	Mg^{2+}
^{1}H-NMR (400 MHz, D_2O, 30°C)						
Chemical shift in ppm: α CH		3.98	4.00	3.68	4.18	4.08
β CH_2		1.98, 1.80	1.99, 1.80	1.68, 1.48	2.16, 1.93	2.05, 1.88
γ CH_2		2.19	2.19	1.93	2.38	2.31
^{13}C-NMR (67.9 MHz, D_2O, 30°C)						
Chemical shift in ppm: α CH		56.43	62.21		62.21	62.10
β CH_2		31.61	35.16		36.17	35.11
γ CH_2		34.01	39.74		39.68	39.60
CO		182.21	182.11		182.16	182.12
COO^-		182.69	185.46		185.82	185.16
FT-IR absorption (KBr), cm^{-1}						
C=O, stretch	1739					
Amide I, N-H bending		1643		1643	1622	1654
Amide II, stretch		1585				
C=O, symmetric stretch	1454	1402		1395	1412	1411
C-N, stretch	1162	1131		1139	1116	1089
N-H, oop bending	698	707		685	669	616
O-H, stretch	3449	3436		3443	3415	3402
Thermal analysis:						
Hydrated water (%)	0	10	42		20	40
Dehydration temp. (C)		109	139		110	122
Melting point (C)	206	160	193, 238	219		160
Decomposition temp. (C)	209.8	340	341	223	335.7	331.8

[a] Molecular weight (kDa) of H-form=1.23 x 10^6, Na-form=1.23 x 10^6, K-form=0.98 x 10^6, NH4-form=0.89 x 10^6, Ca-form=0.49 x 10^6, Mg-form=0.89 x 10^6.

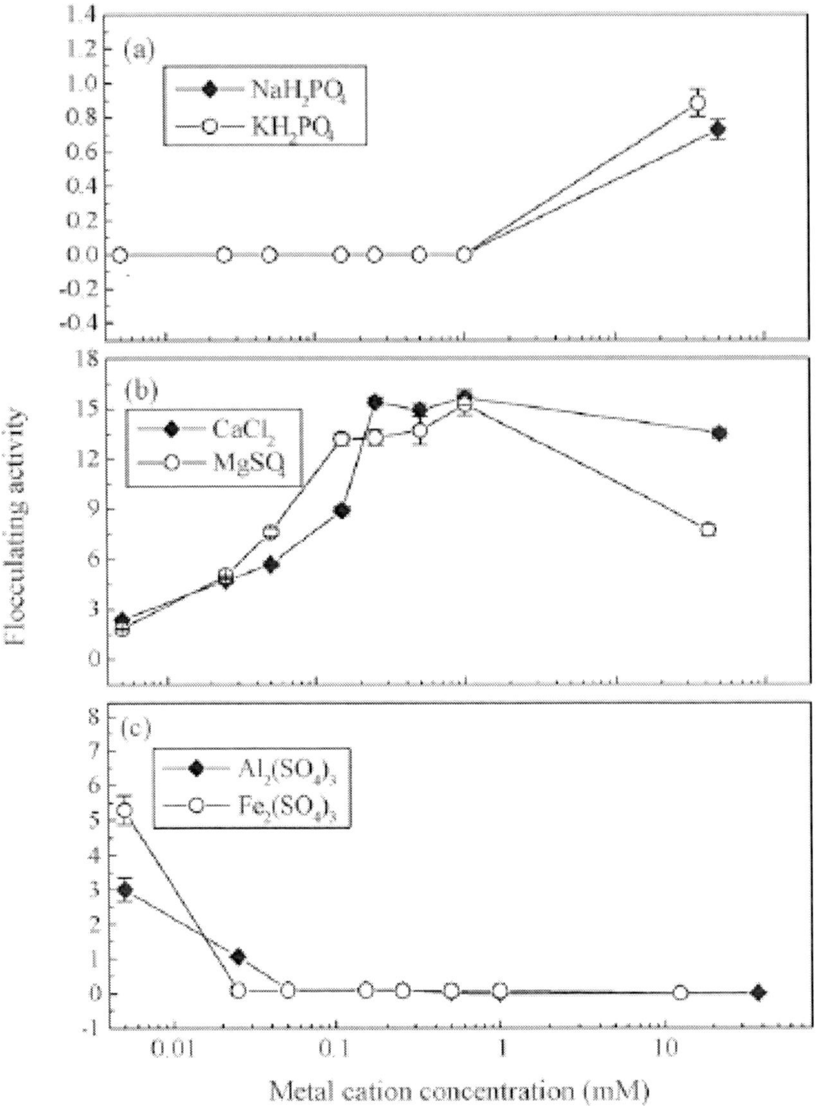

Source: Wsu and Ye 2007.

Figure 4. Effect of metal cation concentration on flocculation activity of γ-PGA from B. subtilis in kaolin suspension. (a) monovalent cations; (b) divalent cations; (c) trivalent cations.

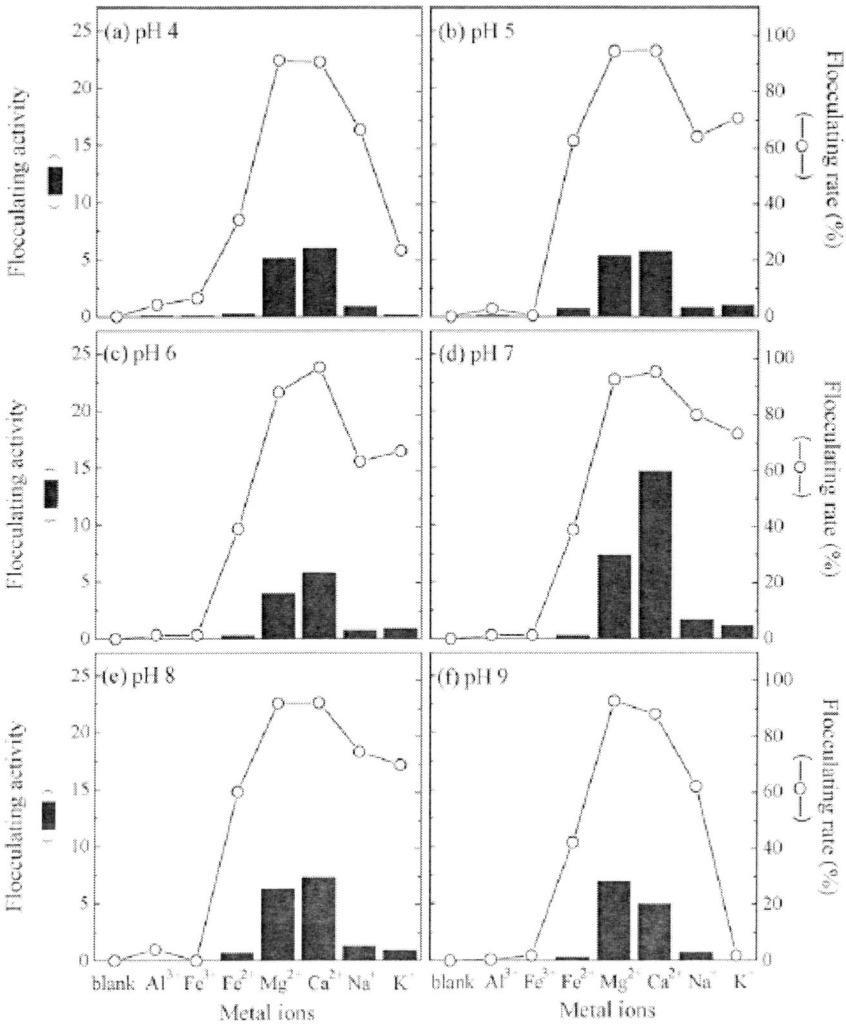

Source: Wu and Ye 2007.

Figure 5. Effect of metal ions and pH on flocculation activity and flocculating rate of kaolin suspension by γ-PGA from B. subtilis (DYU1). Metal concentration (mM): l2(SO4)3=37.5; Fe2(SO4)3=12.5; FeSO4=32.9; CaCO3=49.9; NaH2PO4=32; KH2PO4= 36.7.

Table 3. Flocculation activity of γ-PGA in various inorganic and organic suspensions[a]

Suspensions	Metal cations	Flocculation activity (l/OD)
	Inorganic suspension	
Kaolin[b]	90 mM Ca^{2+}	8.5
Active carbon[c]	8 mM Ca^{2+}	3.3
	2 mM Mg^{2+}	2.9
	0.05 mM Fe^{2+}	2.8
Solid soil[c]	8 mM Ca^{2+}	1.1
	8 mM Mg^{2+}	1.0
	6 mM Fe^{2+}	1.9
Acid clay[c]	6 mM Ca^{2+}	10.9
	4 mM Mg^{2+}	14.0
	0.1 mM Fe^{2+}	34.7
$Ca(OH)_2$[c]	Nil	21.3
$CaCO_3$[c]	Nil	2.0
$[Ca(PO4)_2]_3 \cdot Ca(OH)_2$[c]	Nil	5.7
$Mg(OH)_2$[b]	Nil	8.1
$MgCO_3$[c]	Nil	9.3
Al_2O_3[b]	Nil	6.1
	Organic suspension[b]	
Cellulose powder	90 mM Ca^{2+}	3.0
Carboxymethylcellulose	90 mM Ca^{2+}	2.6
Saccharomyces cerevisiae	90 mM Ca^{2+}	4.6
Bacillus circulans	90 mM Ca^{2+}	2.8

[a] suspension concentration: 5 g/L; [b] treated with γ-PGA from *B. licheniformis* (CCRC 12826) at 80 mg/L; [c] treated with γ-PGA from *B. subtilis* (IFO 3335) at 20, 10 and 10 mg/L for Ca^{2+}, Mg^{2+} and Fe^{2+}, respectively.

Source: Yokoi et al. 1996; Shih et al. 2001.

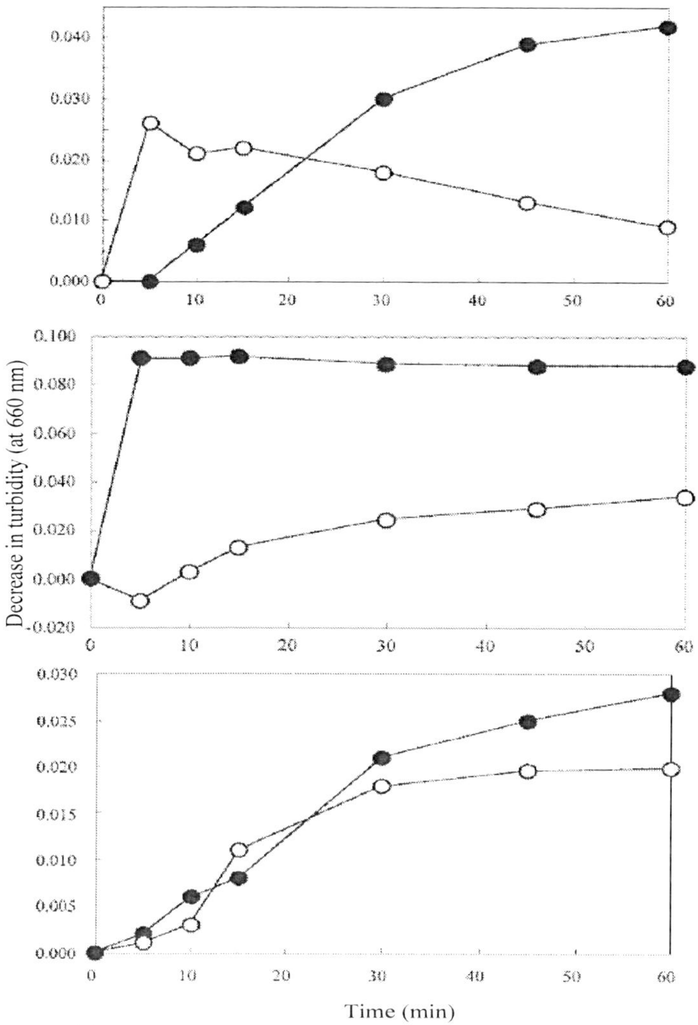

Source: Taniguchi et al. 2005.

Figure 6. Changes in the turbidity of various suspensions by treatment with PAC (incubation at 30°C and pH 5 for 10 min) followed by CL-γ-PGA. (A) M. aeruginosa M21 (closed circles) and E. coli IFO 3992 (open circles) pretreated with PAC (0.5 μg/mL) followed by CL-γ-PGA at 5 and 10 μg/mL, respectively. (B) Bentonite (closed circles) and diatomaceous earth (open circles) pretreated with 2 and 0.5 μg/mL PAC followed by CL-γ-PGA at 5 and 10 μg/mL, respectively. (C) Real sample suspensions, Mandai pond (closed circles) and Yamato river (open circles) pretreated with 2 and 0 μg/mL PAC, respectively, followed by CL-γ-PGA at 20 μg/mL.

3.7. Mechanism of Flocculation

Among various factors, the metal cations played a prominent role in affecting the flocculation activity of γ-PGA. In accordance with Schulze Hardy's law [59], although both suspended particles (like kaolin) and the γ-PGA-based-bioflocculants are anionic, their interaction to facilitate flocculation is possible only through charge neutralization and bridging by added metal cations. Based on this phenomenon, two flocculation mechanisms were suggested [31] as follows:

I. The metal cations may decrease the negative charge on suspended particles, which in turn result in a charge reversal from negative to positive. Finally, the negatively charged carboxyl group (COO^-) of γ-PGA may react with the positively charged site of the suspended cation-kaolin particles (Figure 7a).

II. The metal cations may neutralize and stabilize the residual negative charges of the γ-PGA first, followed by bridge formation with the negatively charged kaolin particles (Figure 7b).

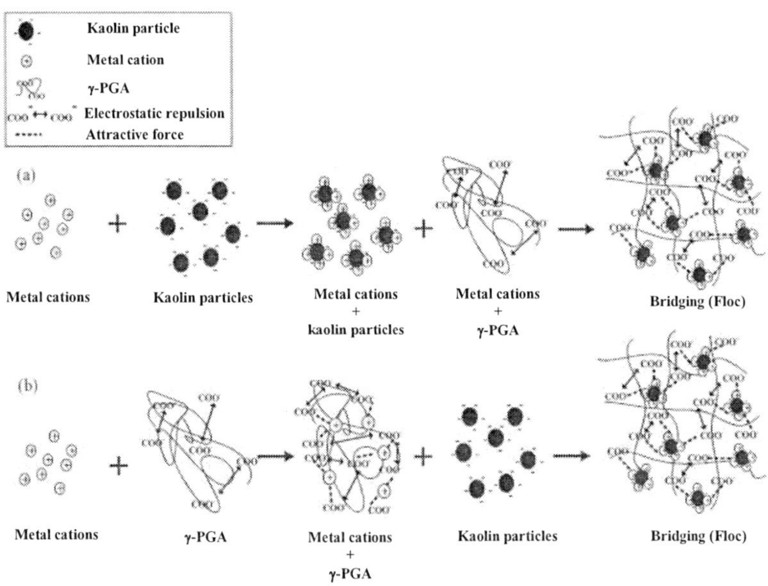

Source: Wu and Ye 2007.

Figure 7. Proposed mechanism of flocculation by γ-PGA.

Thus, metals cations could facilitate charge neutralization and bridging between γ-PGA and suspended particles, resulting in an increase in floc density, floc size and floc resistance to shear. This explains why both the trivalent and bivalent cations possess stronger synergistic effect than monovalent ions.

Table 4. Flocculation activity of cross-linked γ-PGA (CL-γ-PGA) on various suspensions pretreated with different concentrations of polyacrylamide (PAC)

Suspension	CL-γ-PGA (μg/mL)	Flocculation activity (l/OD)[a]						
		PAC (μg/mL)						
		0	0.1 (0.01)[b]	0.25 (0.025)[b]	0.5 (0.05)[b]	1 (0.1)[b]	2 (0.25)[b]	4 (0.5)[b]
Kaolin	10	–	–	–	14.54	39.94	41.56	–
Bentonite	5	0.6	–	–	-5.2	2.9	33.4	0.0
Diatomaceous earth	10	-0.2	–	23.3	30.0	29.6	33.4	–
Escherichia coli	10	2.4	-7.2	16.0	25.9	8.5	1.1	–
Mycrocystis aeruginosa	5	-0.3	6.6	9.5	29.3	41.0	71.7	88.2
Crystal violet	5	60.7	50.4	38.1	18.4	8.1	2.2	0.6
Mandai pond	20	10.8	7.4	7.9	6.1	12.6	23.0	–
Yamato river	20	28.2	25.1	28.7	26.6	26.7	26.1	–

Source: Taniguchi et al. 2005.
[a] incubation at 30° C and pH 5 for 30 min; [b] PAC dose for *Mycrocystis aeruginosa*.

4. γ-PGA as an Adsorbent of Cationic Dyes

Cationic dyes are the brightest class of soluble dyes used by the textile industry and the waste discharge accumulates heavy organic load in the environment causing a harmful impact on both aquatic organisms and human [60-64]. Because of the limitations prevailing in the conventional treatment methods [62], there has been a continuous search for a safe, eco-friendly and biodegradable adsorbent to treat dye-containing wastewaters. Lately, the bioremediation of dyes employing extracellular polymeric substances (EPS) produced by microorganisms is gaining attention [65-69], and this part of the chapter consolidates the published reports on evaluating γ-PGA as an adsorbent of cationic dyes. In all the three studies reported here [27, 70, 71], the γ-PGA (H-form; MW 990 kDa) produced from *B. subtilis* by salvage bioconversion pathway was employed. Figure 8 presents the molecular structure of cationic dyes studied along with their color index name, color index classification number, formula weight and wavelength of maximum absorption (λ_{max}).

4.1. Adsorption Kinetics

Process performance and ultimate cost of an adsorption system depend on the effectiveness of process design and efficiency of process operation, which requires an understanding of the kinetics of uptake or the time dependence of the concentration distribution of the solute in both bulk solution and solid adsorbent. The time (min)-uptake (mg/g) profile for adsorption of 100 or 200 mg/L of cationic dyes by γ-PGA was characterized by a fast dye uptake, attaining equilibrium within 30 or 60 min, respectively, for Au-O, Rh-B and Sa-O dyes. However, a large portion (>85%) of the equilibrium uptake was accomplished within 12 or 20 min, indicating a rapid surface adsorption of cationic dyes did occur on γ-PGA [27]. The kinetic data were modeled with the following pseudo first order and pseudo second order equations [72-74] given by

$$q_t = q_e[1 - \exp(-k_1 t)] \tag{3}$$

$$q_t = \frac{t}{\frac{1}{k_2 q_e^2} + \frac{t}{q_e}} \tag{4}$$

where, q_t and q_e is the dye uptake (mg/g) at time t and equilibrium, respectively, k_1 (L/min) and k_2 (g/mg min) are first order and second order rate constants, and $k_2 q_e^2$ (mg/g min) represents initial adsorption rate (h). Fitted kinetic parameters obtained by non-linear regression (Marquart-Levenberg algorithm) are summarized in Table 5 [27, 70, 71]. Based on two error parameters, r^2 (coefficient of determination) and χ^2 (chi-square statistic), the pseudo second order model gave a more precise fit, suggesting the rate of dye uptake may be largely controlled by chemisorption mechanism involving valence forces through sharing or exchange of electrons [74]. Following a rise in dye concentration from 100-200 mg/L, the second order rate (g/mg min) of dye uptake dropped from 1.04 x 10^{-2} – 4.75 x 10^{-3} for Au-O, 6.95 x 10^{-3} – 4.50 x 10^{-3} for Rh-B and 2.18 x 10^{-3} – 5.25 x 10^{-4} for Sa-O. It is apparent that once the readily available active sites on γ-PGA are occupied, the rate of dye uptake decreases [27].

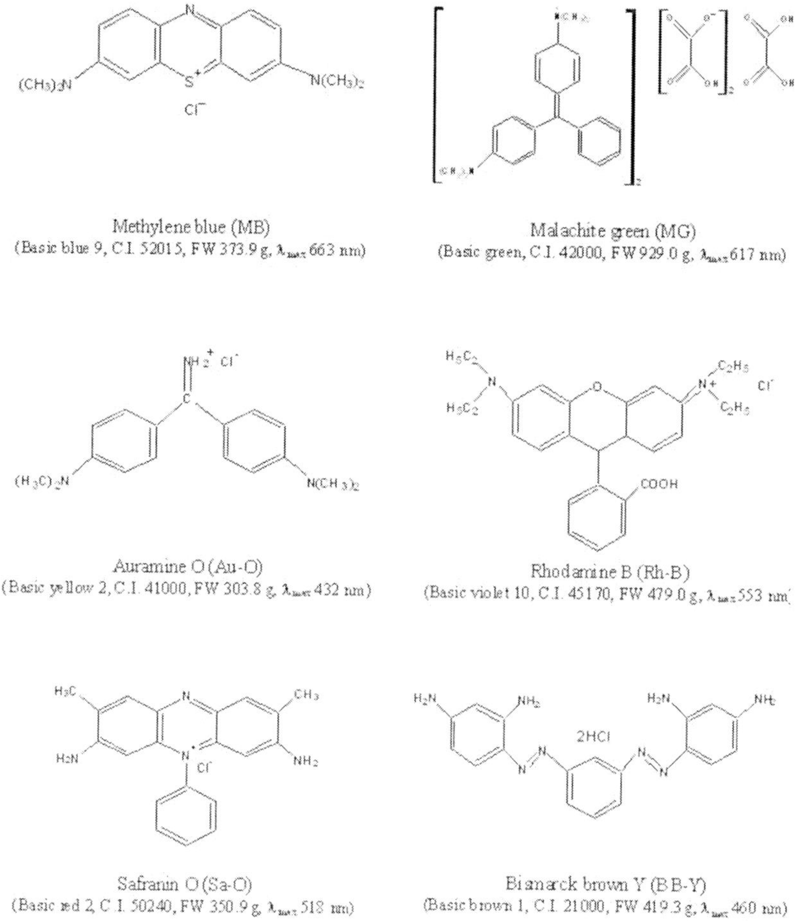

Figure 8. Molecular structure of cationic dyes. C.I.: Color Index classification number; FW: Formula weight.

The adsorption kinetics of BB-Y dye (100 mg/L) conducted at three different temperatures (301, 318 and 333 K) showed a rapid uptake with time (92-98% within 20 min), but diminished on elevating the temperature [71]. The pseudo second order model fitted the kinetic data best, with the derived q_e values being declined by 59.58 mg/g for a temperature raised from 301 to 333

K, accompanied by the rate climbing from 1.50 x 10^{-3} – 6.25 x 10^{-3} g/mg min, probably because of acceleration in the mobility of dye cations at higher temperatures. In addition, the dominant surface adsorption through ion exchange and absence of any particle diffusion may account for the reduced dye uptake at an elevated temperature, as the particle diffusion is an endothermic process [74]. Furthermore, the higher the temperature, the lower the ion exchange capacity and the larger the physical adsorption capacity [75]. The activation energy (E_a, kJ/mol) was determined using the Arrhenius equation [71], which relates the rate constant (k) and temperature (T, Kelvin) according to the expression,

$$k = A\, e^{-\left(E_a/RT\right)} \qquad (5)$$

where A (g/mg min) is the temperature-independent pre-exponential factor. The E_a value obtained was 37.21 kJ/mol, which was greater than the range (<30.00 kJ/mol) reported for a diffusion-controlled reaction [71], implying the adsorption of BB-Y dye by γ-PGA was only reaction-controlled.

Likewise, the kinetic data at different pH (2.00, 3.00, 3.41, 3.94 and 5.00) also fitted the pseudo second order model and the adsorption rate was characterized by a sharp rise from 5.49 x 10^{-4} – 2.48 x 10^{-3} g/mg min for a pH change from 2-3, accounting for a rapid decrease in competing hydrogen ions by 10 times [71]. However, a sudden drop in rate from 2.48 x 10^{-3} – 1.40 x 10^{-3} g/mg min occurred for a pH increase from 3-3.41. When the pH was raised from 2-3 and 3-3.94, the dye uptake increased proportionately by 97.72 and 101.14 mg/g for a 10-fold decrease in the hydrogen ion concentration, respectively [71].

4.1.1. Boyd's Ion Exchange Model

The kinetic data was also analyzed using an ion exchange model proposed by Boyd et al. [78], who postulated the adsorption kinetics for exchange of ions by organic zeolites to be a chemical phenomenon. The exchange reaction between two monovalent ions can be expressed according to the mass law as

$$A^+ + BR \rightleftharpoons B^+ + AR$$

If m_{A+} and m_{B+} denote the concentrations of the ions A^+ and B^+ in solutions, and n_{AR} and n_{BR} the moles of A^+ and B^+ in the adsorbent, respectively, then the net reaction rate can be written as follows:

$$\frac{dn_{AR}}{dt} = k_1(m_{A^+})(n_{BR}) - k_2(m_{B^+})(n_{AR})$$
$$= -n_{AR}(k_1 m_{A^+} + k_2 m_{B^+}) + k_1 m_{A^+} E \quad (6)$$

where k_1 and k_2 are the forward and reverse specific rate constants, respectively, and E is a constant defined by $E = n_{AR} + n_{BR}$. When both concentrations of A^+ and B^+ in solution are kept constant, then, on integration, Equation (6) becomes

$$n_{AR} = \frac{k_1 m_{A^+} E}{k_1 m_{A^+} + k_2 m_{B^+}}(1 - e^{-St}) = q_t \quad (7)$$

where $S = k_1 m_{A^+} + k_2 m_{B^+}$, and Equation (7) can be rewritten as

$$\log(1 - F) = -\left(\frac{S}{2.303}\right)t \quad (8)$$

where F is the fractional attainment of equilibrium at time t, calculated by the ratio between the amounts adsorbed (mg/g) at time t and at infinite time ($F=q_t/q\infty$), and S (L/min) is the ion exchange rate constant. The kinetic data fitted with Equation 8 gave linear curves (plots not shown) with high correlation (r^2=0.954-0.998) and the rate constant (S) values are summarized in Table 6 [27, 70, 71]. The difference in S values for different dye adsorption by γ-PGA may be due to variation in the structural features that affect their mass transfer rate from bulk solution onto γ-PGA.

4.1.2. Mass Transport Mechanism

An adsorption process involves three consecutive mass transport steps such as (i) film diffusion, (ii) intraparticle or pore diffusion and (iii) adsorption onto interior sites [70]. Among these three steps, the last step is considered negligible as it is assumed to be rapid and hence the rate of adsorption should be controlled by either film or particle diffusion. From the practical point of view, it is essential to determine the rate-determining step of an adsorption process. Therefore, the kinetic data were further analyzed using a particle-controlled diffusion expression proposed by Boyd et al. [78, 79], which is given below

$$F = 1 - \frac{6}{\pi^2}\sum_{n=1}^{\infty}\frac{1}{n^2}\exp[-n^2Bt] \quad (9)$$

$$B = \pi 2Di/ro2 \quad (10)$$

where r_o (cm) denotes the radius of the γ-PGA particle determined by sieve analysis, D_i (cm^2/s) is the effective diffusion coefficient and B is the time constant. The Bt values calculated for each value of F were plotted against time t and the plots obtained were shown to be linear (plots not shown), but did not pass through the origin, confirming that film diffusion or external mass transport mainly governs the uptake of cationic dyes by γ-PGA [70]. This tendency also suggested that intraparticle resistance is negligible during the initial period of adsorption. Assuming γ-PGA particles to be spherical in nature, the effective diffusion coefficient (D_i, cm^2/s) was determined using the Equation 10 to be 1.54 x 10^{-4} and 1.55 x 10^{-4} for MB and MG dye adsorption, respectively. Similar D_i values in the order of 10^{-4} were also reported for film-diffusion-controlled adsorption of MB dye by mango seed kernel powder [80]. It was further reported that the film diffusion controls adsorption process in a system with poor mixing, low solute concentration, small adsorbent size and high solute/adsorbent interaction, whereas the opposite is true for the particle diffusion-controlled adsorption process [80]. Apparently, the small-sized γ-PGA particles (1-150 μm) employed for dye adsorption with poor mixing (120 rpm), and the fast dye uptake rate implicating high dye/γ-PGA interaction ascertain that the ion exchange reaction through film diffusion is the rate-controlling process for cationic dye adsorption by γ-PGA.

4.2. Adsorption Isotherms

Adsorption isotherms are the basic requirements for designing an adsorption system as an isotherm expresses the relation between the liquid phase solute concentration and the mass of solute adsorbed at constant temperature per unit mass of adsorbent. Isotherm curves obtained at three different temperatures (301, 318 and 333 k) and at natural pH (without adjusting pH) were characterized by an increase in dye adsorption following a raise in dye concentration from 10-200 mg/L [27, 70, 71]. Conversely, the amount of dye adsorbed declined with a rise in temperature from 301-333 K, implying an exothermic nature of the adsorption process. The shape of isotherm curves resembled either L- or H- type according to classification by Giles et al. [81]. These isotherm types, commonly referred as Langmuir type, signified a high degree of adsorption at very low dye concentration, and suggested the uptake of dyes by γ-PGA to be associated with chemical forces rather than physical interaction [82, 83].

An accurate mathematical description of equilibrium adsorption data is crucial for reliable prediction of adsorption parameters and quantitative comparison of adsorption behavior. The process variables and the underlying thermodynamic assumptions of the equilibrium models often provide insight into the surface property and affinity of the adsorbent as well as the adsorption mechanism. The equilibrium adsorption curves of Au-O, Rh-B, Sa-O and BB-Y by γ-PGA were fitted with three most commonly used isotherm models, namely, Freundlich [84], Langmuir [83] and Redlich-Peterson [85], represented as

$$q_e = K_F C_e^{1/n} \tag{11}$$

$$q_e = \frac{q_m K_e C_e}{1 + K_e C_e} \tag{12}$$

$$q_e = \frac{A C_e}{1 + K_R C_e^{\gamma}} \tag{13}$$

where C_e is the dye concentration in solution at equilibrium, K_F (mg/g) and n are Freundlich constants denoting adsorption capacity and intensity of adsorption, respectively, whereas q_m (mg/g) and K_e (L/mg) represent the maximum adsorption capacity and energy of adsorption. Non-linear regression analysis along with comparison of error parameters (r^2 and χ^2) revealed that the Redlich-Peterson model could describe the equilibrium data more closely than the other two models (Table 7) [27, 71]. The maximum adsorption capacity q_m derived from the Langmuir model was higher or comparable to those reported for nonconventional adsorbents (Table 8) [86-111], demonstrating γ-PGA could be an effective adsorbent in scavenging cationic dyes. The main characteristics of the Langmuir isotherm can be expressed by a dimensionless separation factor or equilibrium parameter R_L, which is defined as $R_L=1/(1+K_eC_o)$, where K_e (L/mg) is the Langmuir constant and C_o (mg/L) is the initial dye concentration. Hall et al. [112] proposed four idealized types of equilibrium behavior, namely, irreversible ($R_L=0$), unfavorable ($R_L>1$), linear ($R_L=1$) and favorable ($0 < R_L < 1$). The R_L values for the dye concentration range 10-200 mg/L in Table 9 were 0-1, indicating a favorable adsorption of cationic dyes by γ-PGA [27, 70, 71].

4.3. Effect of Temperature

By using the equilibrium data obtained for a range of dye concentration at different temperatures (301, 318 and 333 K), several thermodynamic parameters such as change in free energy (ΔG), enthalpy (ΔH) and entropy (ΔS) could be determined using the following equations and are summarized in Table 10 [27, 71].

$$\Delta G^\circ = -RT \ln K_e \tag{14}$$

$$\ln K_e = -\frac{\Delta H^\circ}{RT} + \frac{\Delta S^\circ}{RT} \tag{15}$$

where K_e is the equilibrium constant (Langmuir constant), R is the universal gas constant (8.314 J/mol K) and T is the absolute temperature (K). The negative ΔG° values indicated the spontaneity of dye adsorption, while the negative ΔH° value confirmed the exothermic nature of adsorption process [27,71]. Likewise, the negative ΔS° revealed the decreased randomness at the solid/solution interface and no structural modification occurred in γ-PGA [27,71]. Thus, the negative ΔH°, ΔS° and ΔG° values affirmed that the adsorption of cationic dyes Au-O, Rh-B, Sa-O and BB-Y on γ-PGA was favored at lower temperatures.

4.4. γ-PGA Dose-activity Relationship

On increasing the γ-PGA dose from 0.04-6.00 g/L for a 200 mg/L dye solution, the removal of Au-O, Rh-B and Sa-O dyes rose by 79.1, 81.8 and

67.9%, respectively, while the amount of dye adsorbed declined by 889.9, 800.0 and 1531.7 mg/g [27]. This contrasting behavior may be attributed to a less proportionate increase in dye adsorption following a rise in γ-PGA dose, and reduction in the vicinity of active sites because of aggregation of γ-PGA particles, which in turn may result in removal of some weakly bound dyes from the γ-PGA surface [60,113]. The relationship between γ-PGA dose (m, g/L) and percentage removal (R) fitted the mathematical Equations (16)-(18) with r2 being 0.990, 0.988 and 0.978 for Au-O, Rh-B and Sa-O dyes, respectively [27].

$$R = \frac{m}{(3.39\times10^{-3})+(9.57\times10^{-3}m)} \quad (16)$$

$$R = \frac{m}{(2.37\times10^{-3})+(9.30\times10^{-3}m)} \quad (17)$$

$$R = \frac{m}{(1.13\times10^{-3})+(9.05\times10^{-3}m)} \quad (18)$$

4.5. Effect of pH

Solution pH is a critical parameter which dictates the uptake and release of solutes in any adsorption process involving ion exchange mechanism. Following an increment in pH from 1-5 in a 100 mg/L dye solution, the removal of MB, MG, Au-O, Rh-B, Sa-O and BB-Y dyes rose from 6.7-94.7, 4.0-91.8, 7.6-69.1, 14.5-79.0, 23.0-95.6 and 15.0-93.5%, respectively, and remained constant thereafter [27, 70, 71]. The reduced uptake at low pH may be due to the excessive hydrogen ions in solution competing for active sites in γ-PGA. However, at pH $\geq$ pK_a of γ-PGA (4.09), the carboxyl groups would be largely deprotonated, facilitating maximum exchange of dye cations. A significant drop in solution pH substantiated the concomitant release of hydrogen ions during dye adsorption. For example, the pH of Au-O, Rh-B and Sa-O solutions declined to an average value of 3.86, 3.75 and 3.68, respectively, when the dye solutions at pH 7, 6, 5 and 4 were equilibrated with γ-PGA [27].

4.6. Effect of Electrolytes

Incorporation of different electrolytes (NaCl, KCl, $CaCl_2$ or $MgCl_2$) at varying concentrations (0.001-1 M) showed a declining trend on Au-O, Rh-B and Sa-O dyes adsorption by γ-PGA (Figure 9) [27]. The metal salts added may screen the electrostatic interaction between anionic γ-PGA and cationic dye, and an increase in salt concentration could reduce the amount of dye adsorbed on γ-PGA [114, 115]. The degree of screening by added metal salts followed the order: $Ca^{2+} > Mg^{2+} > K^{+} > Na^{+}$, which may be accounted for by the size and valence of added metal ions [114, 115]. Accordingly, the divalent ions Ca^{2+} and Mg^{2+} showed a more pronounced effect than the monovalent ions Na^{+} and K^{+}, with Ca^{2+} ions exerting a larger effect than Mg^{2+}, and K^{+} ions than Na^{+}. Apparently, the larger the ions, the lower the degree of hydration and the less hydrated ions are drawn closer to the charged surface due to their smaller radii.

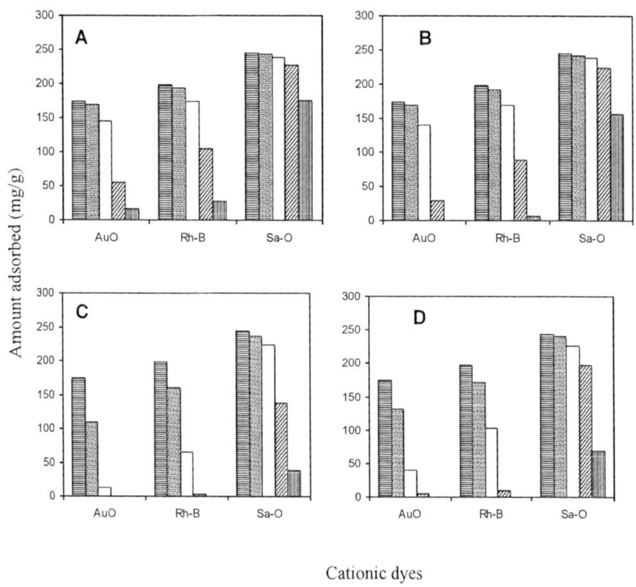

0 M (control); 0.001 M; 0.01 M; 0.1 M; 1 M

Source: Inbaraj et al. 2006a; Table data converted into bar diagram.

Figure 9. Effect of electrolytes at different concentrations on removal of cationic dyes by γ-PGA. Electrolyte type: (A) NaCl; (B) KCl; (C) $CaCl_2$; (D) $MgCl_2$; Electrolyte concentration:

Table 5. Kinetic model parameters for adsorption of cationic dyes by γ-PGA at different experimental conditions

Experimental conditions	Pseudo first order model				Pseudo second order model			
	q_e (mg/g)	k_1 (L/min)	r^2	$\chi^{2\,a}$	q_e (mg/g)	k_2 (g/mg min)	r^2	$\chi^{2\,a}$
Au-O at 100 mg/L [b]	164.54	0.9917	0.914	2.13	172.03	1.04×10^{-2}	0.996	0.09
Au-O at 200 mg/L [b]	249.99	0.7618	0.941	3.41	261.88	4.75×10^{-3}	0.992	0.66
Rh-B at 100 mg/L [b]	184.95	0.8590	0.853	5.32	195.94	6.95×10^{-3}	0.981	0.67
Rh-B at 200 mg/L [b]	295.69	0.5988	0.810	21.23	315.89	4.50×10^{-3}	0.962	4.39
Sa-O at 100 mg/L [b]	226.22	0.3851	0.931	15.42	249.14	2.18×10^{-3}	0.985	2.64
Sa-O at 200 mg/L [b]	414.97	0.1593	0.964	55.38	453.77	5.25×10^{-4}	0.993	5.64
MB at 100 mg/L [c]	226.17	0.8995	0.848	6.45	237.22	6.37×10^{-3}	0.981	0.80
MG at 100 mg/L [c]	185.45	0.9227	0.882	3.68	195.25	7.91×10^{-3}	0.992	0.25
BB-Y at 301 K [d]	267.83	0.2902	0.882	57.49	289.74	1.50×10^{-3}	0.961	15.17
BB-Y at 318 K [d]	233.57	0.5493	0.789	24.91	248.17	3.26×10^{-3}	0.950	6.19
BB-Y at 333 K [d]	221.27	0.8312	0.838	8.66	230.16	6.25×10^{-3}	0.968	1.62
BB-Y at pH 2.00 [e]	87.38	0.0539	0.994	0.74	104.54	5.49×10^{-4}	0.994	1.87
BB-Y at pH 3.00 [e]	177.41	0.3308	0.942	10.26	192.00	2.48×10^{-3}	0.993	1.10
BB-Y at pH 3.41 [e]	248.08	0.2642	0.923	31.20	270.19	1.40×10^{-3}	0.987	4.51
BB-Y at pH 3.94 [e]	267.83	0.2902	0.882	57.49	289.74	1.50×10^{-3}	0.961	15.17
BB-Y at pH 5.00 [e]	285.30	0.2854	0.872	63.63	309.52	1.36×10^{-3}	0.960	16.67

Source: Inbaraj et al. 2006a; 2006c; 2008.

[a] $\chi 2 = \Sigma((q_{ex}-q_{th})^2/q_{th})$, where q_{ex} and q_{th} are experimental and theoretical amounts of dye adsorbed (mg/g) at time t; [b]γ-PGA dose: 0.4 g/L, Temperature: 301 K, pH: natural pH; [c]γ-PGA dose: 0.4 g/L, Temperature: 301 K, pH: natural pH; [d]BB-Y concentration: 100 mg/L, γ-PGA dose: 0.3 g/L, pH: natural pH; eBB-Y concentration: 100 mg/L, γ-PGA dose: 0.3 g/L, Temperature: 301 K, pH: natural pH.

Table 6. Ion exchange rate constants (S) for cationic dye adsorption by γ-PGA at different experimental conditions

Experimental conditions	Boyd's ion exchange model	
	S (L/min)	r^2
Au-O at 100 mg/L[a]	0.2538	0.958
Au-O at 200 mg/L[a]	0.2294	0.934
Rh-B at 100 mg/L[a]	0.1087	0.965
Rh-B at 200 mg/L[a]	0.0864	0.976
Sa-O at 100 mg/L[a]	0.1331	0.967
Sa-O at 200 mg/L[a]	0.0769	0.974
MB at 100 mg/L[b]	0.1918	0.997
MG at 100 mg/L[b]	0.1799	0.988
BB-Y at 301 K[c]	0.1025	0.998
BB-Y at 318 K[c]	0.1082	0.992
BB-Y at 333 K[c]	0.2199	0.987
BB-Y at pH 2.00[d]	0.0548	0.998
BB-Y at pH 3.00[d]	0.1027	0.975
BB-Y at pH 3.41[d]	0.0979	0.985
BB-Y at pH 3.94[d]	0.1025	0.998
BB-Y at pH 5.00[d]	0.0910	0.998

Source: Inbaraj et al. 2006a; 2006c; 2008.

[a] γ-PGA dose: 0.4 g/L, Temperature: 301 K, pH: natural pH; [b] γ-PGA dose: 0.4 g/L, Temperature: 301 K, pH: natural pH; [c] BB-Y concentration: 100 mg/L, γ-PGA dose: 0.3 g/L, pH: natural pH; [d] BB-Y concentration: 100 mg/L, γ-PGA dose: 0.3 g/L, Temperature: 301 K, pH: natural pH.

4.7. Recovery of Adsorbed Dyes

Maximum recovery (>98%) of cationic dyes (MB, MG, Au-O, Rh-B, Sa-O and BB-Y) from dye-treated-γ-PGA could be achieved at pH 1 or 1.5, but tended to decline on raising the solution pH [27, 70, 71]. This desorption trend was opposite to that observed for dye adsorption, suggesting an ion exchange mechanism may be predominantly involved in adsorption of cationic dyes on γ-PGA. This outcome enables the reuse of γ-PGA by subjecting the dye-treated-γ-PGA to dye desorption and adopting proper regeneration procedures without losing its physico-chemical characteristics.

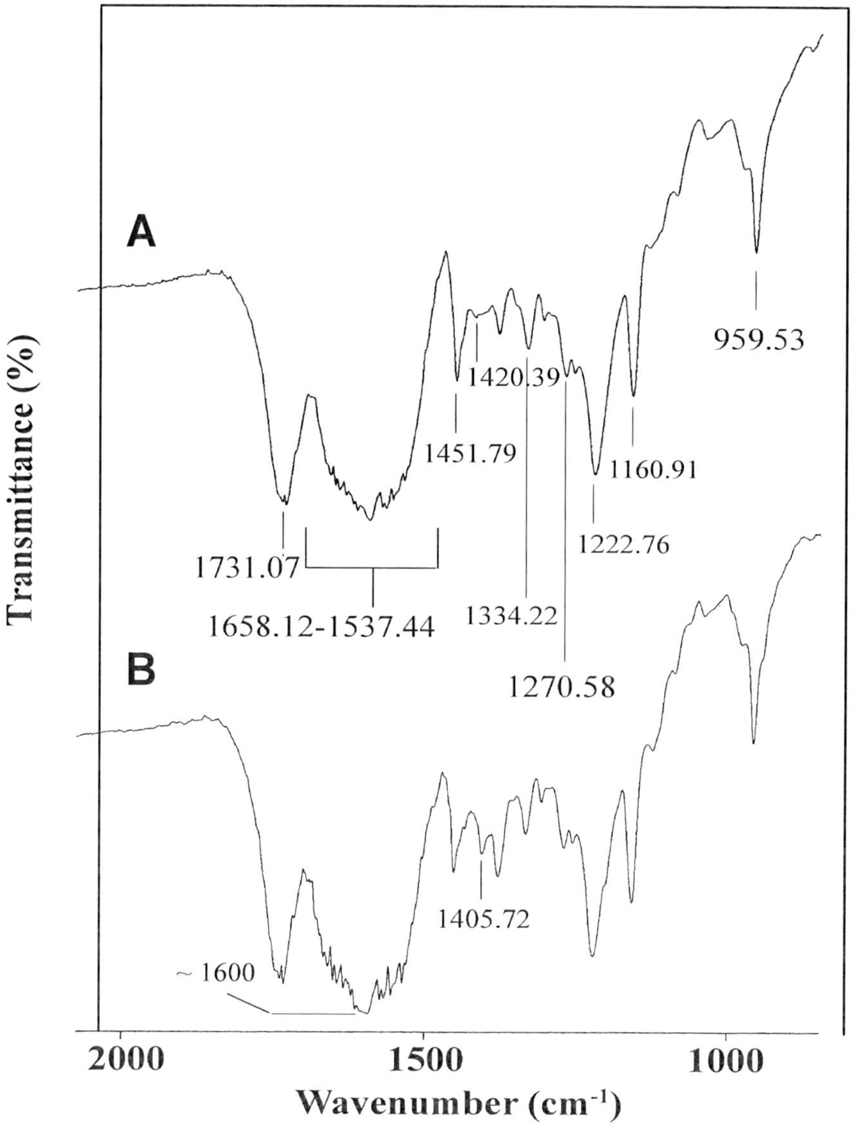

Figure 10. Infra-Red spectra of untreated γ-PGA (A) and BB-Y dye-treated γ-PGA (B).

Table 7. Isotherm model parameters for adsorption of cationic dyes by γ-PGA at different temperaturesa

Isotherm model	Fitted isotherm parameters											
	Au-O/γ-PGA			Rh-B/γ-PGA			Sa-O/γ-PGA			BB-Y/γ-PGA		
	301 K	318 K	333 K	301 K	318 K	333 K	301 K	318 K	333 K	301 K	318 K	333 K
Freundlich												
K_F ($L^n mg^{1-n}/g$)	44.24	27.85	24.30	31.89	15.68	11.02	129.60	95.31	76.84	112.59	68.08	54.42
N	2.54	2.33	2.44	1.79	1.54	1.52	2.33	2.44	2.56	2.82	2.52	2.53
r^2	0.994	0.997	0.999	0.975	0.987	0.978	0.954	0.984	0.975	0.959	0.968	0.973
χ^2	6.33	3.43	0.52	30.68	18.21	20.55	77.84	32.61	33.69	45.58	28.61	18.77
Langmuir												
q_m (mg/g)	277.29	203.60	197.50	390.25	378.00	305.00	502.83	432.26	369.29	469.36	411.57	357.48
K_e (L/mg)	0.07	0.05	0.04	0.04	0.02	0.01	0.30	0.19	0.14	0.17	0.08	0.06
r^2	0.982	0.979	0.967	0.998	0.998	0.995	0.996	0.979	0.988	0.984	0.987	0.991
χ^2	38.79	21.07	44.34	2.84	4.26	6.34	7.04	47.18	25.02	16.23	10.28	5.28
Redlich-Peterson												
A (L/g)	67.32	41.23	259.32	15.61	7.56	7.56	169.51	190.86	88.98	111.41	43.16	29.22
K_R $(L/mg)^\gamma$	1.00	1.00	10.18	0.04	0.02	0.02	0.42	1.18	0.52	0.40	0.20	0.16
γ	0.70	0.65	0.60	1.00	1.00	1.00	0.93	0.73	0.81	0.87	0.86	0.86
r^2	0.999	0.999	0.999	0.998	0.998	0.995	0.997	0.996	0.997	0.992	0.992	0.996
χ^2	0.58	0.44	0.36	2.84	4.26	6.34	8.12	11.40	5.93	11.87	8.91	3.97

Source: Inbaraj et al. 2006a; 2008.

[a] Dye concentration range: 10-200 mg/L; γ-PGA dose: 0.4 g/L for Au-O, Rh-B, Sa-O and 0.3 g/L for BB-Y.

Table 8. Maximum adsorption capacities (q_m) of poly (γ-glutamic acid) compared with those reported for cationic dye removal by some non-conventional adsorbents

Adsorbent	Cationic dye	q_m (mg/g)	Reference
Bark	Safranin O	1119.00	[86]
Bark	Methylene blue	914.58	[86]
Cotton waste	Safranin O	875.00	[86]
Rice husk	Safranin O	838.00	[86]
Corn cob	Safranin O	790.00	[87]
Poly (γ-glutamic acid)[a]	Bismarck brown Y[a]	667.09[a]	[71]
Pinewood	Methylene blue	556.00	[88]
Sugar industry mud	Basic red 22	519.00	[89]
Rice husk carbon	Malachite green	511.00	[90]
Poly (γ-glutamic acid)	Safranin O	502.83	[27]
Poly (γ-glutamic acid)[b]	Bismarck brown Y[b]	469.36[b]	[71]
Tree fern	Basic red 13	408.00	[91]
Peat	Basic violet 14	400.00	[92]
Poly (γ-glutamic acid)	Rhodamine B	390.25	[27]
Waste newspaper	Methylene blue	390.00	[93]
Poly (γ-glutamic acid)	Methylene blue	352.76	[70]
Treated peat	Malachite green	350.00	[92]
Coal	Methylene blue	323.68	[86]
Palm-fruit bunch	Basic yellow	320.00	[94]
Rice husk	Methylene blue	312.00	[86]
Clay	Methylene blue	300.00	[95]
Poly-γ-glutamic acid	Malachite green	293.32	[70]
Poly (γ-glutamic acid)	Auramine O	277.29	[27]
Cotton waste	Methylene blue	277.78	[86]
Activated sludge biomass	Methylene blue	256.41	[96]
Palm-fruit bunch	Basic red 18	242.00	[97]
TriSyl silicas	Malachite green	208.00	[98]
Coir pith	Rhodamine B	203.00	[99]
Activated furniture	Methylene blue	200.00	[100]
Jackfruit peel carbon	Malachite green	166.37	[101]
Hair	Methylene blue	158.23	[86]
Mango seed kernel powder	Methylene blue	153.85	[80]
Spirodela polyrrhiza biomass	Methylene blue	144.93	[102]
Activated tyres	Methylene blue	130.00	[100]
Activated sewage char	Methylene blue	120.00	[100]
Pithophora species	Malachite green	117.65	[103]

Table 8. (Continued)

Adsorbent	Cationic dye	q_m (mg/g)	Reference
Jackfruit peel carbon	Rhodamine B	104.17	[60]
Pearl millet husk carbon	Methylene blue	82.37	[104]
Raw date pits	Methylene blue	80.30	[105]
Pyrolysed furniture	Methylene blue	80.00	[100]
Treated sawdust	Malachite green	74.50	[106]
Modified cyclodextrin	Methylene blue	56.50	[107]
Banana peel	Methylene blue	20.80	[108]
Iron humate	Malachite green	19.20	[109]
Orange peel	Methylene blue	18.60	[108]
Activated date pits (900℃)	Methylene blue	17.30	[105]
Sugarcane dust	Malachite green	4.88	[110]
Neem sawdust	Malachite green	3.42	[111]

[a] At natural pH 4.06 (average pH of concentration range 20-200 mg/L); [b] At pH 5.00.

Table 9. Separation factor (R_L) for removal of cationic dyes by γ-PGA at 301 K

Initial dye concentration (mg/L)	Separation faction (R_L)					
	MB / γ-PGA	MG / γ-PGA	Au-O / γ-PGA	Rh-B / γ-PGA	Sa-O / γ-PGA	BB-Y / γ-PGA
10	0.1354	0.4838	0.5865	0.7415	0.2488	0.2241
20	0.0726	0.3191	0.4149	0.5892	0.1421	0.1615
30	0.0496	0.2380	0.3210	0.4888	0.0994	0.1262
40	0.0377	0.1898	0.2617	0.4177	0.0765	0.1036
50	0.0304	0.1579	0.2210	0.3646	0.0621	0.0878
60	0.0254	0.1351	0.1912	0.3235	0.0523	0.0674
80	0.0192	0.1049	0.1506	0.2640	0.0398	0.0546
100	0.0154	0.0857	0.1242	0.2229	0.0321	0.0459
120	0.0129	0.0724	0.1057	0.1930	0.0269	0.0396
140	0.0111	0.0627	0.0920	0.1701	0.0231	0.0349
160	0.0097	0.0553	0.0814	0.1520	0.0203	0.0281
200	0.0078	0.0448	0.0662	0.1255	0.0163	0.2241

Source: Inbaraj et al. 2006a; 2006c; 2008.

[a] γ-PGA dose: 0.4 g/L for Au-O, Rh-B, Sa-O and 0.3 g/L for BB-Y; Temperature: 301 K.

Table 10. Thermodynamic parameters for adsorption of cationic dyes by γ-PGA[a]

Temperature	K_e (L/mg)	ΔG° (kJ/mol)	ΔH° (kJ/mol)	ΔS° (J/mol K)
		Au-O/γ-PGA		
301	0.071	-7.69	-15.17	-25.13
318	0.046	-6.97		
333	0.040	-6.92		
		Rh-B/γ-PGA		
301	0.035	-7.06	-28.24	-70.85
318	0.016	-5.39		
333	0.012	-4.84		
		Sa-O/γ-PGA		
301	0.302	-11.67	-20.23	-28.46
318	0.194	-11.16		
333	0.139	-10.76		
		BB-Y/γ-PGA		
301	0.173	-10.72	-27.26	-55.45
318	0.079	-9.25		
333	0.062	-9.01		

Source: Inbaraj et al. 2006a; 2008.

[a] Dye conc. range: 10-200 mg/L; γ-PGA dose: 0.4 g/L for Au-O, Rh-B, Sa-O and 0.3 g/L for BB-Y.

4.8. Infra-red (IR) Spectra of Untreated and Dye-treated γ-PGA

The IR spectra of both untreated and dye-treated γ-PGA were almost identical (Figure 10) [71]. Nevertheless, by critical comparison the presence of two new peaks at ~1600 and 1405.72 cm^{-1} in dye-treated γ-PGA could be observed, which are characteristic of asymmetric and symmetric stretching vibrations of carboxylate anion (COO^-), respectively [71]. Thus, the adsorption of cationic dyes on γ-PGA may occur by interaction with carboxylate anions through exchange of hydrogen ions from the side chain carboxyl groups.

5. γ-PGA As an Adsorbent of Chemical Mutagens

Many polysaccharides and dietary fibers from vegetables, fruits and cereals are known to be antimutagenic [116-121]. Nevertheless, the antimutagenic activity of extracellular polymeric substances derived from bacterial sources remains less explored. Recently, the adsorption and antimutagenic behavior of γ-PGA for several chemical mutagens, listed in Figure 11, have been investigated [28, 32] and are discussed herein.

5.1. Binding of Mutagenic Heterocyclic Amines at Gastrointestinal pH

Inbaraj et al. [28] investigated the adsorption characteristics of three mutagenic heterocyclic amines (HA), namely, MeIQ, 4,8-DiMeIQx and Trp-p-2, by γ-PGA from *B. subtilis* (H-form; MW 990 kDa) at gastrointestinal pH conditions. Heterocyclic amines were monitored by a HPLC method developed by Chen and Yang [122], who employed a binary solvent system of acetonitrile and 0.05 ammonium acetate solution (pH 3.6) with the following gradient elution: 9% A at 0 min, 15% A at 8 min, 27% at 18 min, 55% A at 28 min and 100% A at 30 min. A batch-mode equilibrium study at different solution pH (1-7) showed appreciable binding of all HAs over the entire gastrointestinal pH range typically being 2.5 (stomach), 7.5 (small intestine), 5.5 (cecum/ascending colon) and 6.8 (distal colon) (Figure 12). Binding curves developed at pH 2.5 and 5.5 exhibited different isotherm shapes belonging to S and L types, respectively (plots not shown), according to classification by Giles et al. [81]. The S-type curve at pH 2.5 was characterized by an initial concave portion signifying less or no adsorption, which may be attributed to clustering or agglomeration of HA molecules caused by repulsive force between protonated primary amino group of HA and secondary amino group of γ-PGA. In contrast, at high HA concentration, the interaction between HA and γ-PGA became stronger due to multiple physical interaction resulting in a convex shape. The maximum adsorption capacity (q_m, mg/g) derived by fitting the isotherm data at pH 2.5 with a linear form of the classical Langmuir equation ($1/q_e=1/q_mK_eC_e+1/q_m$) [83] followed the order: Trp-p-2 (1428.57) > MeIQ (1250.00) > 4,8-DiMeIQx (666.67) (Table 11). The difference in q_m values between HAs may be due to variation in their hydrophobic character as reported by Ferguson and Harris [122], who demonstrated the adsorption of nine HAs on dietary fiber α-cellulose to increase with hydrophobicity. Also, the number of bulky methyl groups in HA may be responsible for this effect. Accordingly, 4,8-DiMeIQx containing three methyl groups showed the lowest q_m value (666.67 mg/g), followed by MeIQ (1250.00 mg/g) and Trp-p-2 (1428.57 mg/g). Thus, the interaction between HAs and γ-PGA at pH 2.5 may largely depend on physical forces.

On the other hand, the isotherm curves developed for HA concentrations 100-2000 mg/L at pH 5.5 were characterized by two distinct portions (curve I and curve II) with a curve shift appearing between the same C_o values (600 and 800 mg/L) for all the three HAs (plots not shown). This phenomenon suggested the presence of more than one adsorption site with varied affinities and binding energies for HA molecules. Fitting curve I and II individually with the Langmuir equation ($C_e/q_e=C_e/q_m+1/K_eq_m$) provided good correlation ($r^2 > 0.98$), suggesting saturation of two binding sites with varying binding energies. Linear fitting of curve I yielded the isotherm parameters for initial monolayer saturation, while curve II gave relavant parameters for overall HA binding by γ-PGA (Table 12). Moreover, fitting with the Scatchard equation ($q_e/C_e=q_mK_e-q_eK_e$) [123] showed a significant deviation from linearity, but a good correlation ($r^2 > 0.91$) was observed when applied separately to curves I and II, confirming the presence of at least two different types of binding (Table 12). The multisite HA adsorption at pH 5.5 was also interpreted on the basis of a multimolecular adsorption by using the BET equation [124] as given below:

$$\frac{C_e}{(C_s - C_e)} = \frac{1}{aq_m} + \left(\frac{a-1}{aq_m}\right)\left(\frac{C_e}{C_s}\right) \tag{19}$$

where C_s (mg/l) is the solute concentration at the saturation of all layers, q_m (mg/g) is the amount of HA required to form a unimolecular layer and a is a BET constant representing average heat of adsorption in the first layer. Linear plots of $C_e/(C_s - C_e)$ versus C_e/C_s gave a high correlation for 4,8-DiMeIQx and Trp-p-2 ($r^2 > 0.96$), but not for MeIQ ($r^2 > 0.87$). From the fitted parameters, the q_m value for the unimolecular layer was found to be ~2.5 times higher than that obtained by the Langmuir equation for saturation of all layers (Table 13), implying the multimolecular HA adsorption did occur for at least two layers on γ-PGA. According to the generalized polarization theory by DeBoer and Zwicker [125], the adsorption of non-polar molecules on ionic adsorbents like γ-PGA may induce dipoles in the first layer of adsorbed molecules, which in turn may stimulate dipoles on the second layer favoring a multilayer adsorption. Based on the hydrophobic character of HAs and ionic nature of γ-PGA, the adsorption on first layer may be formed by hydrogen-bonding, hydrophobic and ionic interactions, while the subsequent layer (s) may result from weak dipole-dipole and hydrophobic interactions.

Table 11. Langmuir parameters for heterocyclic amines binding by γ-PGA at pH 2.5[a]

HA	Langmuir parameters		
	q_m (mg/g)	b (l/mg)	r^2
MeIQ	1250.00	0.0020	0.948
4,8-DiMeIQx	666.67	0.0004	0.933
Trp-p-2	1428.57	0.0005	0.952

Source: Inbaraj et al. 2006b.

[a] Experimental conditions: HA concentration range: 10-2000 mg/L; γ-PGA dose: 1 mg/mL.

5.2. Suppressive Effect on SOS Response of Salmonella Typhimurium Induced by Chemical Mutagens

Sato et al. [32] recently reported the suppressive effect of γ-PGA (Na form) from *B. subtilis* on SOS response (a post replication DNA repair system that allow DNA replication to bypass lesions) of *Salmonella typhimurium* (TA1535/pSK1002) induced by direct and indirect forms of chemical mutagens (AF-2, MNNG, 4NQO, Trp-p-2, IQ and MeIQx). The antimutagenicity was determined by a slightly modified *umu* test [126], which was based on the ability of γ-PGA to suppress the *umu* operon expression induced by DNA-damaging chemical mutagens. A plasmid (pSK 1002) carrying fusion gene *umu*C-*lac*Z was introduced into *S. typhimurium* TA 1535 and the percentage of suppression was calculated in terms of the β-galactosidase activity of cells resulting from the expression and suppression of the *umu* operon according to the equation: (1-A/B) x 100, where A and B are the β-galactosidase activity with and without γ-PGA, respectively. The concentration of chemical mutagens used to induce the SOS response in *S. typhimurium* was 0.3, 6, 3, 0.3, 0.3 or 3 μg/mL of AF-2, MNNG, 4NQO, IQ,

MeIQx or Trp-p-2, respectively, which was chosen based on induction of β-galactosidase activity without affecting the cell growth.

Table 12. Langmuir and Scatchard plot parameters for heterocyclic amines binding by γ-PGA at pH 5.5[a]

HA	Curve I			Curve II		
	q_m (mg/g)	b (l/mg)	r^2	q_m (mg/g)	b (l/mg)	r^2
		Langmuir model parameters				
MeIQ	500.00	0.2632	0.989	1428.57	0.0374	0.999
4,8-DiMeIQx	370.37	0.2030	0.991	1000.00	0.0152	0.999
Trp-p-2	500.00	0.6897	0.982	1666.67	0.0211	0.999
		Scatchard plot parameters				
MeIQ	499.16	0.1805	0.918	1465.09	0.0432	0.983
4,8-DiMeIQx	397.79	0.0706	0.994	944.68	0.0169	0.991
Trp-p-2	454.30	1.5903	0.913	1619.74	0.0228	0.995

Source: Inbaraj et al. 2006b.

[a] Experimental conditions: HA concentration range: 10-2000 mg/L; γ-PGA dose: 1 mg/mL.

Table 13. BET isotherm parameters for heterocyclic amines binding by γ-PGA at pH 5.5[a]

HA	BET model parameters		
	q_m (mg/g)	a	r^2
MeIQ	653.60	51.00	0.867
4,8-DiMeIQx	370.37	451.00	0.961
Trp-p-2	623.44	401.00	0.989

Source: Inbaraj et al. 2006b.

[a] Experimental conditions: HA concentration range: 10-2000 mg/L; γ-PGA dose: 1 mg/mL.

The γ-PGA (Na form; 4000 kDa) tested at three different concentrations (1, 2 and 3%) showed a substantial suppressive effect on the SOS response induced by all the mutagens, with >80% suppression being achieved by a 3% solution (Figure 13). Irrespective of the type of chemical mutagen, the antimutagenic activity tested for γ-PGA (3%) with different molecular masses (50-8000 kDa) exhibited a lower effect at 50 kDa and >6000 kDa when compared to others, with maximum activity being shown for γ-PGA at 4000 kDa (Figure 3). The low water-holding capacity of γ-PGA at 50 kDa and the three dimensional structure or the rheological property of γ-PGA at >6000 kDa may be responsible for the less-pronounced effect in entrapping the mutagens [32]. However, the variation in salt form of γ-PGA (Na, K or Ca) did not show any remarkable suppressive effect on the SOS response induced by indirect mutagens (IQ, MeIQx and Trp-p-2), implying the difference in electronegativity or valency of metal ions poses no significant impact. The antimutagenic activity of γ-PGA (3% of 4000 kDa) was also compared with its monomer units, D- and L-glutamic acids (Na form), and some other reference compounds like carboxymethylcellulose (CMC) and xanthan gum (XG) [32]. The percentage of suppression by γ-PGA was the highest for all chemical mutagens, followed by XG, CMC, and D- and L-glutamates (Table 14). Although XG and CMC are similar to γ-PGA in possessing carboxyl groups, the difference in antimutagenic activity was large, probably because of variation in ion exchange capacity (meq/g) between γ-PGA (8.7) and XG (2.0) or CMC (0.68). Besides, the γ-PGA also exhibited significant antimutagenic effect on the SOS response of *S. typhimurium* induced by the active forms (nitrenium ions) of three indirect mutagens (Trp-p-2, IQ and MeIQx) [32].

Table 14. Suppressive effect of γ-PGAa and reference compounds on SOS response of S. typhimurium induced by various chemical mutagens

Mutagen (Final concentration)	Suppression (%)				
	CMC[b]	XG[c]	LG[d]	DG[e]	γ-PGA
AF-2 (0.3 μg/mL)	25	40	15	12	90
MNNG (6 μg/mL)	30	72	20	18	80
4NQO (3 μg/mL)	40	75	20	22	81
IQ (0.3 μg/mL)	20	48	38	43	90
MeIQx (0.3 μg/mL)	35	70	17	25	90
Trp-p-2 (3 μg/mL)	30	43	27	25	82

Source: Sato et al. 2008.

[a] 3% Na-form γ-PGA at 4000 kDa; [b]carboxymethylcellulose; [c]xanthan gum; [d]L-glutamic acid; [e]D-glutamic acid.

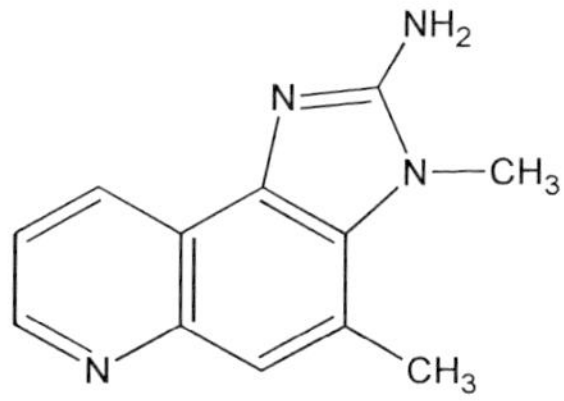

2-amino-3,4-dimethylimidazo[4,5-f] quinoline (MeIQ)

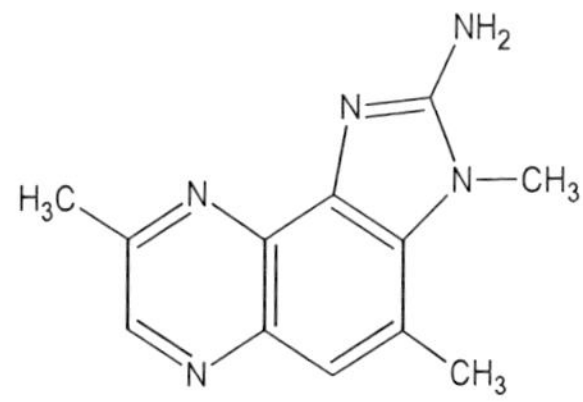

2-amino-3,4,8-trimethylimidazo[4,5-f] quinoxaline (4,8-DiMeIQx)

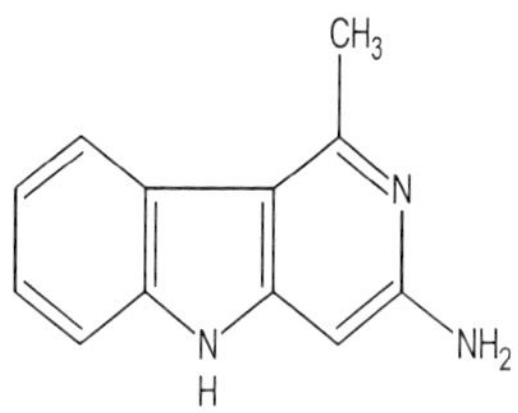

3-amino-1-methyl-5H-pyrido[4,3-b] indole (Trp-p-2)

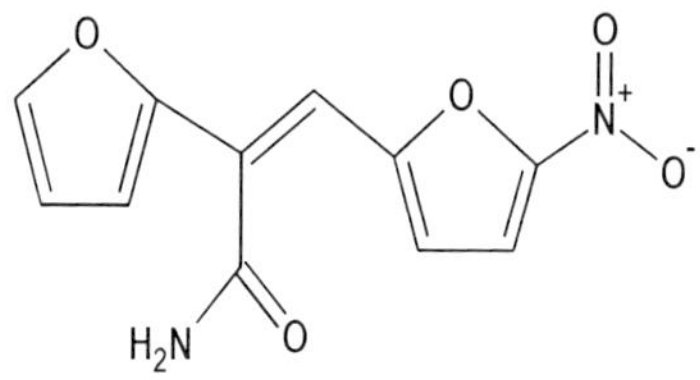

Furylframide (AF-2)

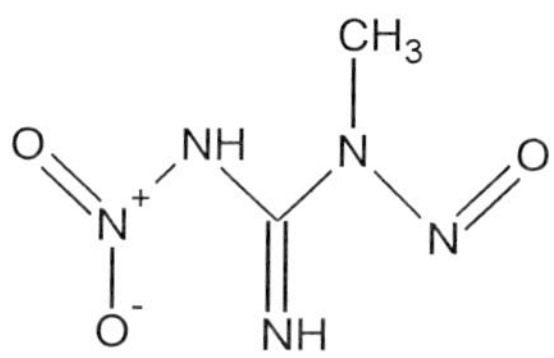

N-methyl-N′-nitro-N-nitrosoguanidine (MNNG)

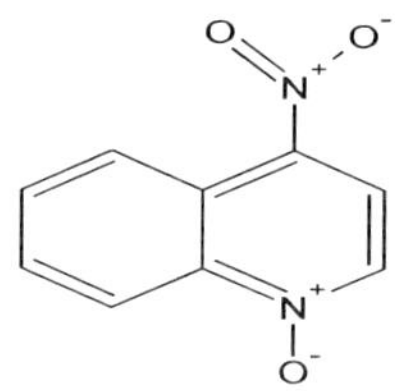

4-nitroquinoline 1-oxide (4NQO)

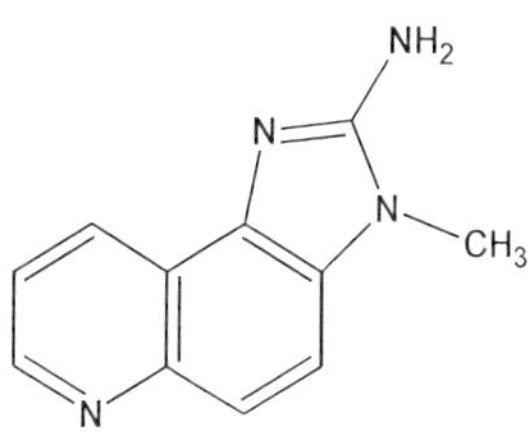

2-amino-3-methylimidazo[4,5-f] quinoline (IQ)

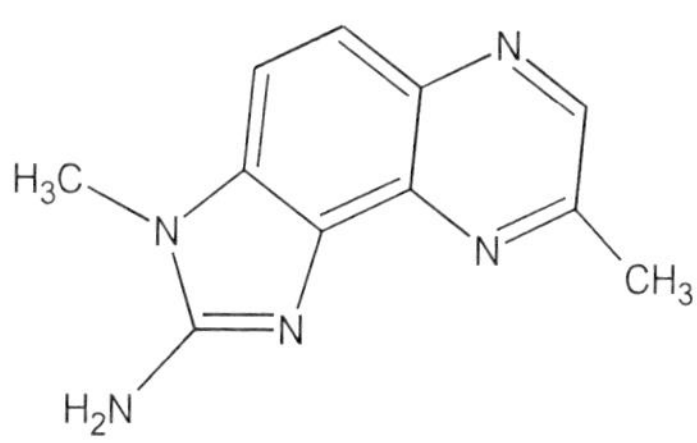

2-amino-3,8-dimethylimidazo[4,5-f] quinoxaline (MeIQx)

Figure 11. Molecular structure of chemical mutagens.

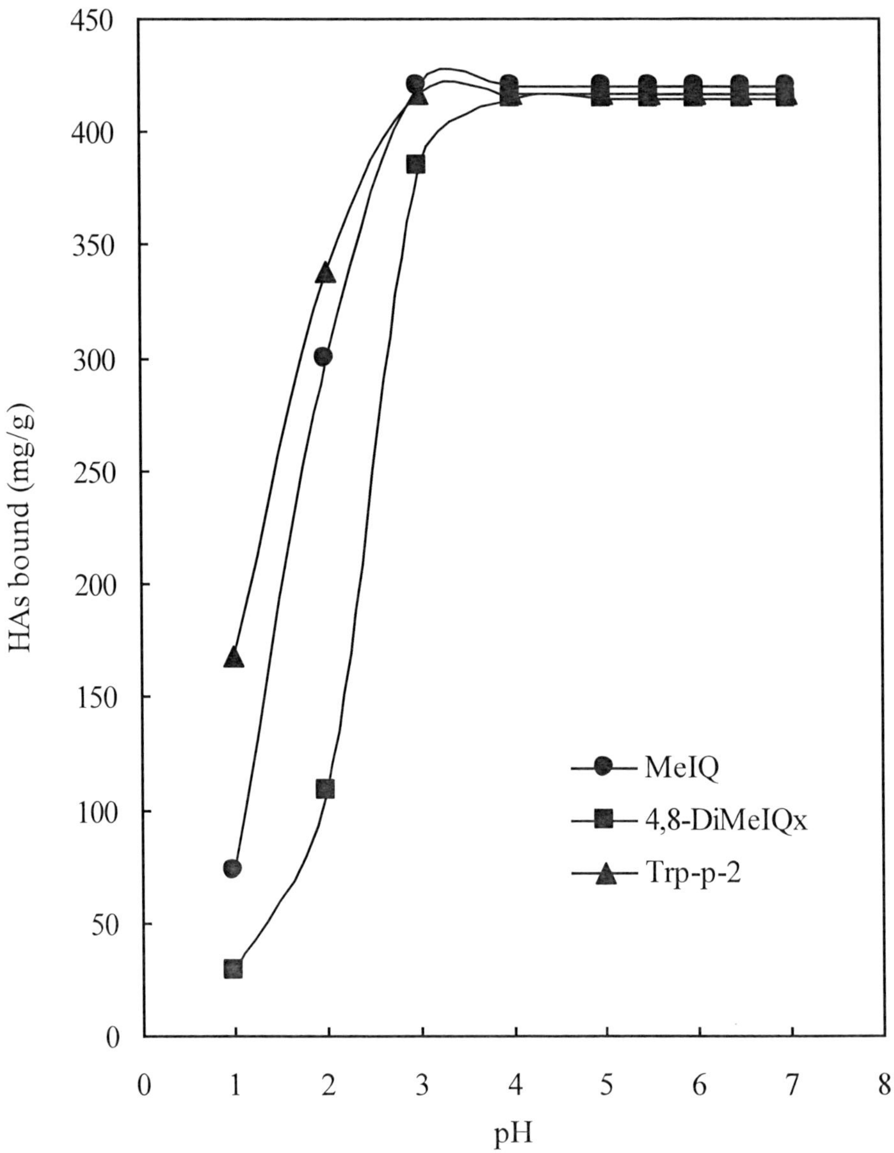

Source: Inbaraj et al. 2006b.

Figure 12. Effect of pH on binding of heterocyclic amines (HA) by γ-PGA. HA concentration: 500 mg/L; γ-PGA dose: 5 mg/mL.

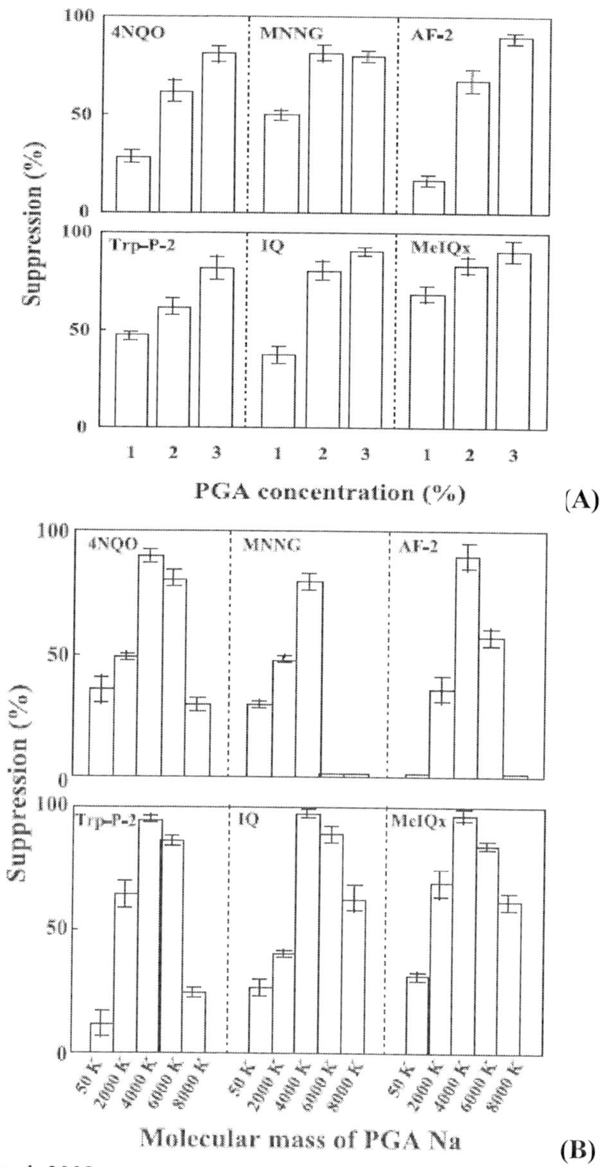

Source: Sato et al. 2008.

Figure 13. Effect of γ-PGA concentration (A) and molecular mass (B) on suppression of SOS response of S. *typhimurium* induced by various chemical mutagens. Mutagen concentration: AF-2=0.3 μg/mL; MNNG=6 μg/mL; 4NQO=3 μg/mL; Trp-p-2=3 μg/mL; IQ=0.3 μg/mL; MeIQx=0.3 μg/mL, γ-PGA: Na-form at 4000 kDa. The vertical error bars represent standard deviation for a mean of triplicate experiments.

Conclusion

The foregoing review has demonstrated the eco-friendly and biocompatible biopolymer poly(γ-glutamic acid) can be a potential material for flocculating turbid waters, and scavenging cationic dyes and chemical mutagens. Various physical and chemical factors affecting the flocculation as well as adsorption of cationic dyes and chemical mutagens were elaborated. Though the synthetic alternatives may be effective and inexpensive, their application could be restricted from the environmental and health point of view. On the contrary, biopolymers are generally non-toxic and biodegradable, but the major pitfall limiting the practical application is their high cost. Nevertheless, the commercial viability may be enhanced by synthesizing extracellular biopolymers through large-scale fermentation by employing simple and cost-effective downstream processing methods. Future research on these topics should focus on a more realistic system involving column operations for cationic dye removal and *in vivo* study for scavenging chemical mutagens.

References

[1] Ivánovics, G.; Bruckner, V. *Z. Immunitatsforsch.* 1937, 90, 304-318.
[2] Ivánovics, G.; Erdös, L. *Z. Immunitatsforsch.* 1937, 90, 5-19.
[3] Fujii, H. *Nippon Nogeikagaku Kaishi*, 1963, 37, 407-411.
[4] Fujii, H. *Nippon Nogeikagaku Kaishi*, 1963, 37, 474.
[5] Shih, I. L.; Van, Y. T. *Bioresour. Technol.* 2001, 79, 207-225
[6] Bovarnick, M. *J. Biol. Chem.* 1942, 145, 415-424.
[7] Ward, R. M.; Anderson, R. F.; Dean, F. K. *Biotechnol. Bioeng.* 1963, 5, 41-48.
[8] Troy, F. A. *J. Biol. Chem.* 1973, 248, 305-316.
[9] Murao, S. *Koubunshi* 1969, 16, 1204-1212.
[10] Thorne, C. B.; Gómez, C. G.; Blind, G. R.; Housewright, R. D. *J. Bacteriol.* 1953, 65, 472-478.
[11] Thorne, C. B.; Gómez, C. G.; Noyes, H. E. Housewright, R. D. *J. Bacteriol.* 1954, 68, 307-315.
[12] Housewright, R. D. In *The Bacteria: A Treatise on Structure and Function*; Gunsalus, I. C.; Stanier, R. Y.; Eds.; Academic Press: New York, 1962; pp 389-412.

[13] Nitccki, D. E.; Goodman, J. W. In *Chemistry and Biochemistry of Amino Acids, Peptides and Proteins*; Weinstein, B.; Ed.; Marcel Dekker: New York, 1971, pp 87-92.
[14] Sung, M. H. ; Park, C.; Kim, C. J.; Poo, H.; Soda, K.; Ashiuchi, M. *Chemical Record* 2005, 5, 352-366.
[15] Ho, G. H.; Ho, T. I.; Hsieh, K. H.; Su, Y. C.; Lin, P. Y.; Yang, J.; Yang, K. H.; Yang, S. C. *J. Chin. Chem. Soc.* 2006, 53, 1363-1384.
[16] Xu, H.; Jiang, M.; Li, H.; Lu, D.; Quyang, P. *Process Biochem.* 2005, 40, 519-523.
[17] Shih, I. L.; Van, Y. T.; Shen, M. H. *Mini-Rev. Med. Chem.* 2004, 4, 179.
[18] Matsusaki, M.; Serizawa, T.; Kishida, A.; Endo, T.; Akashi, M. *Bioconjugate Chem.* 2002, 13, 23.
[19] Mark, S. S.; Crusberg, T. C.; DaCunha, C. M.; Di Iorio, A. A. *Biotechnol. Prog.* 2006, 22, 523.
[20] Radu, J. E. F.; Novak, I.; Hartmann, J. F.; Borbely, *J. Polym. Prep.* 2006, 47, 420.
[21] Kang, H. S.; Park, S. H.; Lee, Y. G.; Son, T. I. *J. Appl. Polym. Sci.* 2007, 103, 386.
[22] Shimokuri, T.; Kaneko, T.; Serizawa, T.; Akashi, M. *Macromol. Biosci.* 2004, 4, 407.
[23] Perez-Camero, G.; Garcia-Alvarez, M.; Mrtinez de Iiarduya, A.; Fernandez, C.; Campos, L.; Munoz-Guerra, S. *Biomacromolecules* 2004, 5, 144.
[24] Matsusaki, M.; Serizawa, T.; Kishida, A.; Akashi, M. *Biomacromolecules* 2005, 6, 400.
[25] Chiu, Y. T.; Chiu, C. P.; Chien, J. T.; Ho, G. H.; Yang, J.; Chen, B. H. *J. Agric. Food Chem.* 2007, 55, 5123-5130.
[26] Kao, T. H.; Chen, B. H. *Molecules*, 2007, 12, 917-931.
[27] Inbaraj, B. S.; Chien, J. T.; Ho, G. H.; Yang, J.; Chen, B. H. *Biochem. Eng. J.* 2006a, 31, 204-215.
[28] Inbaraj, B. S.; Chiu, C. P.; Chiu, Y. T.; Ho, G. H.; Yang, J.; Chen, B. H. *J. Agric. Food Chem.* 2006b, 54, 6452-6459.
[29] Inbaraj, B. S.; Wang, J. S.; Lu, J. F.; Siao, F. Y.; Chen, B. H. *Bioresour. Technol.* 2009, 100, 200-207.
[30] Shih, I. L.; Van, Y. T.; Yeh, L. C.; Lin, H .G.; Chang, Y. N. *Bioresour. Technol.* 2001, 78, 267-272.
[31] Wu, J. Y.; Ye, H. F. *Process Biochem.* 2007, 42, 1114-1123.
[32] Sato, M.; Kanie, K.; Soda, K.; Yokoigawa, K. *J. Biosci. Bioeng.* 2008, 105, 690-693.

[33] Goto, A.; Kunioka, M. *Biosci. Biotechnol. Biochem.* 1992, 56, 1031-1035.
[34] Kubota, H.; Nambu, Y.; Endo, T. *J. Polym. Sci. Part A: Polym. Chem.* 1993, 31, 2877-2878.
[35] Kubota, H.; Matsunobu, T.; Uotani, K.; Takebe, H.; Satoh, A.; Tanaka, T.; Tanguchi, M. *Biosci. Biotechnol. Biochem.* 1993, 57, 1212-1213.
[36] Cromwick, A. M.; Gross, R. A. *Biotechnol. Bioeng.* 1996, 50, 222-227.
[37] Kunioka, M.; Goto, A. *Appl. Microbiol. Biotechnol.* 1994, 40, 867-872.
[38] Ito, Y.; Tanaka, T.; Ohmachi, T.; Asada, Y. *Biosci. Biotechnol. Biochem.* 1996, 60, 1239-1242.
[39] Cheng, C.; Asada, Y.; Aaida, T. *Agric. Biol. Chem.* 1989, 53, 2369-2375.
[40] Ogawa, Y.; Yamaguchi, F.; Yuasa, K.; Tahara, Y. *Biosci. Biotechnol. Biochem.* 1997, 61, 1684-1687.
[41] Kunioka, M. *Appl. Microbiol. Biotechnol.* 1997, 47, 469-475.
[42] Birrer, G. A.; Cromwick, A. M.; Gross, R. A. *Int. J. Biol. Macromol.* 1994, 16, 265-275.
[43] Troy, F. A. In *Peptide Antibiotics: Biosynthesis and Functions*; Kleinkauf, H.; Von Döhren, H.; Eds.; Walter de Gruyter: New York, 1982; pp 49-83.
[44] Crescenzi, V.; Dentini, M.; Mattei, B. *ACS Sump. Ser.* 1996, 627, 233 (Chapter 19).
[45] Zanuy, D.; Aleman, C.; Munoz-Guerra, S. *Intl. J. Biol. Macromol.* 1998, 23,175.
[46] Tong, Z.; Zhe, L.; Huai-Lan, Z. *J. Environ. Sci.* 1999, 11, 1-12.
[47] Kurane, R.; Takeda, K.; Suzuki, T. *Agric. Biol. Chem.* 1986, 50, 2301-2307.
[48] Salehizadeh, H.; Vossoughi, M.; Alemzadeh, I. *Biochem. Eng. J.* 2000, 5, 39-44.
[49] Gutcho, S. *Waste Treatment with Polyelectrolytes and other Flocculants*; Noyes Data Corp.: Park Ridge, NJ, 1977; pp 1-37.
[50] Nakamura, J.; Miyashiro, S.; Hirose, Y. *Agric. Biol. Chem.* 1976, 40, 1341-1347.
[51] Salehizadeh, H.; Shokaosadati, S. A. *Biotechnol. Adv.* 2001, 19, 371-385.
[52] Dearfield, K. L.; Abermathy, C. O. *Mutant. Res.* 1988, 195, 45-77.
[53] Vanhorick, M. Moens, W. *Carcinogenesis* 1983, 4, 1459-1463.
[54] Yokoi, H.; Natsuda, O.; Hirose, J.; Hayashi, S.; Takasaki, Y. *J. Ferment. Bioeng.* 1995, 79, 378-380.

[55] Yokoi, H.; Arima, T.; Hirose, J.; Hayashi, S.; Takasaki, Y. *J. Ferment. Bioeng.* 1996, 82, 84-87.

[56] Takagi, H.; Kadowaki, K. *Chitin Nat. Technol.* 1985, 3, 121-128.

[57] Taniguchi, M.; Kato, K.; Shimauchi, A.; Ping, X.; Fujita, K. I.; Tanaka, T.; Tarui, Y.; Hirasawa, E. *J. Biosci. Bioeng.* 2005, 99, 130-135.

[58] Taniguchi, M.; Kato, K.; Shimauchi, A.; Ping, X.; Nakayama, H.; Fujita, K. I.; Tanaka, T.; Tarui, Y.; Hirasawa, E. *J. Biosci. Bioeng.* 2005, 99, 245-251.

[59] Klute, R.; Neis, U. In *Proceedings of the International Conference, Colloid Interface Science*; 50th Edition; Ker, M.; Ed.; 1976; Vol.4, pp 113.

[60] Inbaraj, B. S.; Sulochana, N. *Indian J. Chem. Technol.* 2006, 13, 17-23.

[61] Pearce, C. I.; Lloyd, J .R.; Guthrie, J. T. *Dyes Pigments* 2003, 58, 179-196.

[62] Blackburn, R. S. *Environ. Sci. Technol.* 2004, 38, 4905-4909.

[63] Choy, K. K. H.; Porter, J. F.; McKay, G. *Langmuir* 2004, 20, 9646-9656.

[64] Crini, G. *Bioresour. Technol.* 2006, 97, 1061-1085.

[65] Wingender, J.; Neu, T. R.; Flemming, H. C. *Microbial Extracellular Polymeric Substances: Characterization, Structures and Function*; Springer-Verlag: Heidelberg, 1999.

[66] Liu, H.; Fang, H. H. P. *Biotechnol. Bioeng.* 2002, 80, 806-811.

[67] Gutnick, D. L.; Bach, H. *Appl. Microbiol. Biotechnol.* 2000, 54, 451-460.

[68] Kazy, S. K.; Sar, P.; Singh, S. P.; Sen, A. K; D'Souza, S. F. *World J. Microbiol. Biotechnol.* 2002, 18, 583-58.

[69] Kim, S. Y.; Kim, J. H.; Kim, C. J.; Oh, D. K. *Biotechnol. Lett.* 1996, 18, 1161-1164.

[70] Inbaraj, B. S.; Chiu, C. P.; Ho, G. H.; Yang, J.; Chen, B. H. *J Hazard. Mater.* 2006c, *B*137, 226-234.

[71] Inbaraj, B. S.; Chiu, C. P.; Ho, G. H.; Yang, J.; Chen, B. H. *Bioresour. Technol.* 2008, 99, 1026-1035.

[72] Lagergren, K. *Sven. Ventenskapsakad. Handl.* 1898, 24, 1-39.

[73] Blanchard, G.; Maunaye, M.; Martin, G. *Water Res.* 1984, 18 1501-1507.

[74] Ho, Y. S.; McKay, G. *Process Biochem.* 2003, 38, 1047-1061.

[75] Vlasov, A. V.; Prokof'ev, K. V.; Sangalov, Y. A.; Kotov, S. V. *Chem. Technol. Fuels Oils*, 1988, 24, 52-55.

[76] Ho, Y. S.; Ng, J. C. Y.; McKay, G. *Separ. Purif. Meth.* 2000, 29, 189-232.
[77] Lazaridis, N. K.; Asouhidou, D. D. *Water Res.* 2003, 37, 2875-2882.
[78] Boyd, G. E.; Adamson, A. W.; Myers Jr. L. S. *J. Am. Chem. Soc.* 1947, 69, 2836-2848.
[79] Reichenberg, D. *J. Am. Chem. Soc.* 1953, 75, 589-597.
[80] Kumar, K. V.; Kumaran, A. *Biochem. Eng. J.* 2005, 27, 83-93.
[81] Giles, C. H.; MacEwan, T. H.; Nakhwa, S. N.; Smith, D. *J. Chem. Soc.* 1960, 111, 3973-3993.
[82] Sumner, M. E. *Handbook of Soil Science*; CRC Press: Boca Raton, FL, 2000.
[83] Langmuir, I. *J. Am. Chem. Soc.* 1918, 40, 1361-1368.
[84] Freundlich, H. M. F. *Z. Phys. Chem.* 1906, 57, 384-470.
[85] Redlich, O.; Peterson, D. L. *J. Phys. Chem.* 1959, 63, 1024-1026.
[86] McKay, G.; Porter, J. F.; Prasad, G. R. *Water Air Soil Pollut.* 1999, 114, 423-438.
[87] Juang, R. S.; Wu, F. C.; Tseng, R. L. *Colloid Surf. A: Physiochem. Eng. Aspect* 2002, 201, 191-199.
[88] Tseng, R. L.; Wu, F. C.; Juang, R. S. *Carbon* 2003, 41, 487-495.
[89] Magdy, Y. H.; Daifullah, A. A. M. *Waste Manage.* 1998, 18, 219-226.
[90] Guo, Y.; Yang, S.; Fu, W.; Qi, J.; Li, R.; Wang Z.; Xu, H. *Dyes Pigments* 2003, 56, 219-229.
[91] Ho, Y. S.; Chiang, T. H.; Hsueh, Y. M. *Process Biochem.* 2005, 40, 119-124.
[92] Sun, Q.; Yang, L. *Water Res.* 2003, 37, 1535-1544.
[93] Okada, K.; Yamamoto, N.; Kameshima, Y.; Yasumori, A. *J. Colloid Interf. Sci.* 2003, 262, 194-199.
[94] Nassar, M. M. *Adsorpt. Sci. Technol.* 1997, 15, 609-617.
[95] Bagane, M.; Guiza, S. *Ann. Chim. Sci. Mater.* 2000, 25, 615-626.
[96] Gulnaz, O.; Kaya, A.; Matyar, F.; Arikan, B. *J. Hazard. Mater.* 2004, 108, 183-188.
[97] Nassar, M. M.; Hamoda, M. F.; Radwan, G. H. *Water Sci. Technol.* 1995, 32, 27-32.
[98] Karadağ, E.; Saraydin, D.; Aydin, F. *Turkish J. Chem.* 1998, 22, 227-236.
[99] Namasivayam, C.; Kumar, M. D.; Selvi, K.; Begum, R. A.; Vanathi, T.; Yamuna, R. T. *Biomass Bioenergy* 2001, 21, 477-483.
[100] Sainz-Diaz, C. I.; Griffiths, A. J. *Fuel* 2000, 79, 1863-1871.

[101] Inbaraj, B. S.; Sulochana, N. *Indian J. Chem. Technol.* 2002, 9, 201-208.

[102] Waranusantigul, P.; Pokethitiyook, P.; Kruatrachue, M.; Upatham, E. S. *Environ. Pollut.* 2003, 125, 385-392.

[103] Kumar, K. V.; Sivanesan, S.; Ramamurthi, V. *Process Biochem.* 2005, 40, 2865-2872.

[104] Inbaraj, B. S.; Selvarani, K.; Sulochana, N. J. Sci. Ind. Res. 2002, 61, 971-978.

[105] Banat, F.; Al-Asheh, S.; Al-Makhadmeh, L. *Process Biochem.* 2003, 39, 193-202.

[106] Garg, V. K.; Gupta, R.; Yadav, A. B.; Kumar, R. *Bioresour. Technol.* 2003, 89, 121-124.

[107] Crini, G.; Peindy, H. N. *Dyes Pigments* 2006, 70, 204-211.

[108] Annadurai, G.; Juang, R. S.; Lee, D. J. *J. Hazard. Mater.* 2002, 92, 263-274.

[109] Janoš, P. *Environ. Sci. Technol.* 2003, 37, 5792-5798.

[110] Ho, Y. S.; Chiu, W. T.; Wang, C. C. *Bioresour. Technol.* 2005, 96, 1285-1291.

[111] Khattri, S. D.; Singh, M. K. *Water Air Soil Pollut.* 2000, 120, 283-294.

[112] Hall, K. R.; Eagleton, L. C.; Acrivos, A.; Vermeulen, T. *Ind. Eng. Chem. Fundam.* 1966, 5, 212-219.

[113] Manohar, D. M.; Anoop Krishnan, K.; Anirudhan, T. S. *Water Res.* 2002, 36, 1609-1619.

[114] International Atomic Energy Agency, *Application of Ion Exchange Processes for the Treatment of Radioactive Waste and Management of Spent Ion Exchangers*; International Atomic Energy Agency: Vienna, Austria, 2002, Technical Report No. 408.

[115] Vermöhlen, K.; Lewandowski, H.; Narres, H. D.; Schwuger, M. J. *Colloids and Surfaces A: Physicochem. Eng. Aspects* 2000, 163, 45-53.

[116] Vikse, R.; Mjelva, B. B.; Klungsøyr, L. *Food Chem. Toxicol.* 1992, 30, 239-246.

[117] Ferguson, L. R.; Roberton, A. M.; Watson, M. E.; Triggs, C. M.; Harris, P. J. *Chem. Biol. Interact.* 1995, 95, 245-255.

[118] Harris, P. J.; Triggs, C. M,; Roberton, A. M.; Watson, M. E.; Ferguson, L. R. *Chem. Biol. Interact.* 1996, 100, 13-25.

[119] Ryden, P.; Roberton, J. A. *Mutat. Res.* 1996, 351, 45-52.

[120] Isobe, Y.; Yokoigawa, K.; Kawai, H. *Food Sci. Technol. Int.* 1998, 4, 77-79.

[121] Barnes, W. S.; Maiello, J.; Weisburger, J. H. *J. Natl. Cancer Inst.* 1988, 70, 757-760.
[122] Chen, B. H.; Yang, D. J. *Chromatographia* 1998, 48, 223-230.
[123] Ferguson, L. R.; Harris, P. J. *Mutat. Res.* 1996, 350, 173-184.
[124] Scatchard, G. *Ann. NY Acad. Sci.* 1949, 51, 660-672.
[125] Brunauer, S.; Emmett, P. H.; Teller, E. *J. Am. Chem. Soc.* 1938, 60, 309-319.
[126] DeBoer, J. H.; Zwicker, C. Z. *Physik. Chem.* 1929, *B*3, 407-418.
[127] Oda, Y.; Nakamura, S.; Oki, I.; Kato, T.; Shinagawa, H. *Mutat. Res.* 1985, 147, 219-229.

In: Glutamic Acid
Editors: D. Balcazar and E. Perez
ISBN: 978-1-62257-236-6

Chapter 6

Biosynthesis of Poly (γ-L-Glutamic Acid) and Comparative Studies on the Biodegradability of the γ- and α- Enantiomeric Forms of the Poly (Glutamic Acid) by *Bacillus Licheniformis* NCIMB 11709[*]

M. Soledad Marqués-Calvo[1,†], Jordi Bou[2] and Marta Cerdà-Cuéllar[3]

[1]Departament d'Òptica i Optometria, EUOOT, Universitat Politècnica de Catalunya, Barcelona, Spain

[2]Departament d'Enginyeria Química, ETSEIB, Universitat Politècnica de Catalunya, Barcelona, Spain

[3]Centre de Recerca en Sanitat Animal (CReSA), Esfera UAB, Campus de Bellaterra, UAB, Barcelona, Spain

[*] A version of this chapter also appears in *Environmental and Regional Air Pollution*, edited by Dean Gallo and Richard Mancini, published by Nova Science Publishers, Inc. It was submitted for appropriate modifications in an effort to encourage wider dissemination of research.

[†] To whom correspondence should be addressed: M. Soledad Marqués-Calvo, Departament d'Òptica i Optometria, Universitat Politècnica de Catalunya, C/Violinista Vellsolà, 37.08222 Terrassa, Spain. vre.marques@upc.edu and marques@oo.upc.edu.

Abstract

Poly-γ-glutamic acid (γ-PGA) is a water soluble, polyanionic, extracellular polymer produced by several members of the genus *Bacillus*, the most notable among them *Bacillus subtillis* and *Bacillus licheniformis.* This polymer consists of D- and L-glutamic acid repeating units that are linked between the α-amino and γ-carboxylic acid functionalities. A peculiar characteristic of γ-PGA formed by *B. licheniformis* is that the polymer stereochemistry can be controlled by the composition of the medium used in the polymer production in the laboratory. γ-PGA is water soluble, biodegradable, edible and non-toxic to humans and the environment. Therefore, the potential applications of this polymer and its derivatives have been of growing interest in the past few years in a broad range of industrial fields such as food, cosmetics, medicine and water treatment.

Most of the reports published on the microbial or enzymatic biodegradation of microbial γ-PGA are only concerned with γ-D-PGA. In the present work, γ-L-PGA was biosynthesized using *B. licheniformis* NCIMB 11709, and the biodegradation of the different enantiomeric forms of PGA, as well as the α and γ forms (γ-L-PGA, γ-D-PGA, γ-D, L-PGA, α-L-PGA and α-D-PGA) of this bacterial strain, were studied. γ-PGA with L-units content of 70% was produced with a Mw of 1 million g/mole. *B. licheniformis* NCIMB 11709 did not grow when the α-forms of the polymer were used as sole carbon source, independently of the D or L contents. However, a decrease in the molecular weight of the α-L-PGA in the bacterial culture was observed, which was not due to hydrolytic degradation. On the other hand, this strain could degrade and use the polymer when it was in the γ-form, regardless of the D or L contents.

This finding can have an implication in defining bacterial factors involved in the degradation of the polymer. Moreover, the generated knowledge of the degradation of the different forms of the poly (glutamic acid) will facilitate selection of the better form according to the intended application.

Introduction

Poly-γ-glutamic acid (γ-PGA) is a naturally-occurring homo-polyamide made of D- and L-glutamic acid units connected by amide linkages between α-amino and γ-carboxylic acid groups (Figure 1). Several microorganisms produce this polymer (Table 1), but it is mainly synthesized by various members of the genus *Bacillus*.

Table 1. Organisms reported to produce PGA

Organism	Configuration	Content (%)		References
		D-Enantiomer	L-Enantiomer	
Bacillus anthracis	D	100	0	Hanby and Rydon (1946) [40]
Bacillus licheniformis	D and L	10–100	0–90	Pérez-Camero et al. (1999) [26]
Bacillusmegaterium	D and L	30	70	Ashiuchi et al. (2003) [19]
Bacillus pumilus	D and L	60	40	Schneerson et al. (2003) [41]
Bacillus subtilis (*chungkookjang*)	D and L	60–70	30–40	Ashiuchi and Misono (2002) [27]
Bacillus subtilis (*natto*)	D and L	50–80	20–50	Ashiushi and Misono (2002) [27]
Bacillus halodurans	L	0	100	Aono (1987) [42]
Planococcus halophilus	D	ND	ND	Kandler et al. (1983) [43]
Sporosarcina halophila	D	ND	ND	Kandler et al. (1983) [43]
Staphylococcus epidermidis	D and L	40	60	Kocianova et al. (2005) [44]
Natrialba aegyptiaca	L	0	100	Hezayen et al. (2001) [45]
Hydra	L	0	100	Weber (1990) [46]

ND, not determined.

The specific biological functions of the polymer are due to its poly (anionic) nature and to its water solubility. This poly (amino acid) is present in many different environments and protects the producing organisms against adverse environmental conditions like dehydration or high salt concentration of the medium, against the effects of heavy metal ions or against the response of the human immune system during infections [1]. The role of PGA depends on whether it is anchored or released to the medium. Thus, anchored PGA is a virulence factor, whereas released PGA is a storage element or a persistence factor.

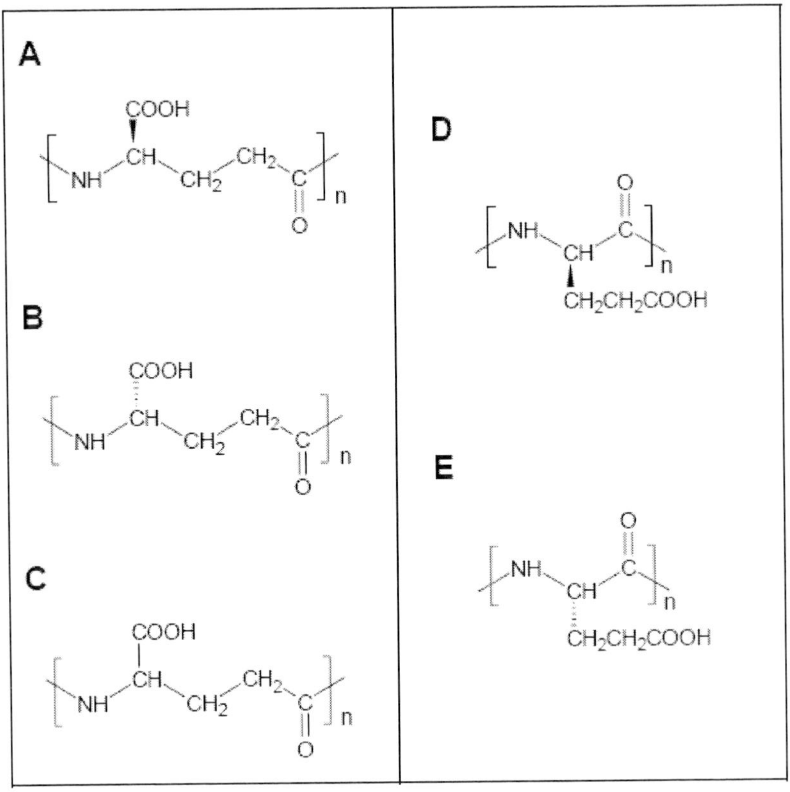

Figure 1. Chemical structures of PGA. A: Poly-γ-D-glutamate (γ-D-PGA), B: poly-γ-L-glutamate (γ-L-PGA), C: poly-γ-DL-glutamate (γ-DL-PGA), D: poly-α-D-glutamate and E: poly-α-L-glutamate. The term *n* indicates the number of glutamyl linkages in PGA and the range of the *n* value is estimated to be 100 to over 10,000.

Because γ-PGA is water soluble, biodegradable, edible and non-toxic to humans and the environment, it could act as a substitute for synthetic polymers with similar characteristics. Therefore, it has been of growing interest in the past few years due to the potential applications of this polymer and its derivatives in a broad range of industrial, medical, pharmaceutical, and personal care applications, as well as in agriculture and environmental fields. γ-PGA and its derivatives are used, among other applications, as a thickener, moisturizer in cosmetics, cryoprotectant, sustained-release material, drug carrier, curable biological adhesive, biodegradable fibre, biopolymer flocculant, heavy metal absorber from water, dispersant, animal feed additive, intestinal Ca^{2+} solubility agent, in antibody production, and as an antitumor agent [2-4]. Biodegradability is one important advantage of naturally-occurring poly (amino acid)s over many synthetic polymers.

The minimal gene sets required for γ-PGA synthesis and release were recently defined. Four genes are involved in the synthesis and only one more is necessary for the release. The synthesis complex is probably responsible for the stereochemical specificity of γ-PGA composition [1]. γ-PGA is optically active and its specific rotation depends on the different proportion of the two enantiomers, L and D.

Polymer production in the laboratory and its stereochemistry can be controlled by the composition of the medium used. Various factors such as inorganic salts, glutamic acid, citric acid, glycerol and size of inoculum affect the production of γ-PGA by *B. licheniformis* in static and shaken cultures. In general, bacteria producing γ-PGA are L-glutamic acid dependent, meaning that exogenous L-glutamic acid is necessary for γ-PGA production. The production of γ-PGA increases with the addition of L-glutamic acid to the medium. Nevertheless, some genetically stable strains of *B. subtilis* and *B. licheniformis* are L-glutamic acid-independent bacteria and can synthesize γ-PGA without the addition of L-glutamic acid to the medium [2]. Poly (glutamic) acid is present as poly (acid) or poly (salt) form, depending on the pH of the aqueous solution.

Molecular weight and polydispersity are important features of microbial PGA for the effect that molecular size has on polymer properties. γ-PGA can have molecular weights ranging from 100 to over 1,000 KDa [5]. Final molecular weight depends on many factors, including medium composition, mode of feeding and fermentation conditions [6-11].

The γ-PGA produced by *B. licheniformis* ATCC 9945a is a homopolymer of D- and L-glutamic acid repeat units. The proportion of D-glutamic acid grows gradually as the concentration of Mn^{+2} increases. Synthesis and assembly of bacterial poly (γ–D-glutamic acid) occur through the function of a membranous multienzyme complex, or polyglutamyl synthetase complex, acting in a concerted fashion [12]. This complex catalyzes the activation, racemization, and polymerization of L-glutamic acid to form a high molecular weight polymer of γ–D–glutamic acid [13]. The extracellular synthetase enzyme activity was measured in *B. licheniformis* but was not detectable in cultures of *B. subtilis* [14]. Ashiuchi et al. [15] showed evidence that a multienzyme system, PgsBCA, was the sole machinery of γ-PGA synthesis in *B. subtilis (chungkookjang)*.

Few reports exist on the microbial and enzymatic biodegradation of microbial γ-PGA. Already in the early studies with *B. licheniformis* a depolymerase degrading γ-PGA was described [16]. Later, several authors have described enzymes from *B. licheniformis* and *B. subtilis* that can hydrolize the γ-glutamyl linkage. These studies deal with the kind of cleavage that takes place (between L- and L-, D- and D-, or D- and L- glutamic acid units) and if the enzyme acted in and *endo*- or *exo*-type manner [17-22]. Also, Oppermann et al. [23] investigated the ability of a non-adapted microbial sewage sludge community to grow with γ-D-PGA as sole carbon and nitrogen source, and several bacteria were isolated and identified. However, the existing studies of PGA degradation by *B. licheniformis* are mainly concerned only with γ-D-PGA and none of them compare the degradation of the different enantiomeric forms of PGA, and its α and γ forms. The present study is concerned with the ability of *B. licheniformis* NCIMB 11709 to degrade all of them. The biosynthesis of γ-L-PGA was also carried out as it was not possible to obtain the commercialized product.

At present, our studies are focussed to develop new block copolymers with controlled-structure in which the bacterial γ-PGA is one of the components. These block copolymers have a clear industrial applicability. To deepen the knowledge of the biodegradation of the different forms of γ-PGA will facilitate the selection of the most suitable form of the polymer for the synthesis of the block copolymers, according to the physical and the chemical characteristics of the final polymer, and to the biodegradation response.

Bacteria and Polymers

The bacteria *B. licheniformis*NCIMB 11709 (ATCC 9945a) from the National Collection of Industrial and Marine Bacteria (Great Britain) was used to synthesize the poly (γ-L-glutamic acid) and to study its ability to degrade the different forms of PGA. The following powdered enantiomeric forms of poly (glutamic acid) were used in this study: i) poly (γ-70%L-glutamic acid), produced in the present study by *Bacillus licheniformis* NCIMB 11709 as described below; ii) poly (γ-77%D-glutamic acid), previously synthesized in our laboratory from *Bacillus licheniformis* NCIMB 11709; iii) poly (γ-D, L-glutamic acid), kindly provided by Dr. Kubota of Meiji Co. (Japan); iv) poly (α-D-glutamic acid) and poly (α-L-glutamic acid), purchased from Sigma–Aldrich, Inc. (Madrid, Spain) and used without further purification.

Polymer Biosynthesis and Characterization

Highly mucoid colonies of the bacterium grown on modified Sauton's agar (Thorne et al. [24]) were selected and grown overnight aerobically at 37°C in a reciprocally shake flask containing 10 ml of modified Sauton's medium (Table 2). Two ml of this culture was used to inoculate two 1L Erlenmeyer flasks each containing 250 ml of the fermentation medium, which was medium E (Table 3) without $MnSO_4$. The pH of the medium was adjusted to 6.5 since it is the optimum for a maximum production of γ-PGA [8]. The cultures were incubated at 37°C in shake flasks at 250 rpm, for 74 h. Studies from Cromwick and Gross [7] and Pérez-Camero et al. [26] demonstrated that both the yield and the enantiomeric content of γ-PGA obtained in cultures of *B. licheniformis* were modulated by the concentration of Mn^{2+} in the culture medium. The L-glutamic acid in γ-PGA varied inversely with the content in $MnSO_4$ over a range of 80% to 10% for $[Mn^{2+}]$ increasing from 0 to 1230 $\mu mol \cdot L^{-1}$. The γ-PGA yield varied from 1 to 6 g L^{-1} over such an interval of manganese concentrations. Since we wanted to obtain a γ-PGA with the maximum content of L-glutamic acid units, we used medium E with no $MnSO_4$, although the yield was going to be low. Optimal γ-PGA production has been determined for both *B. subtilis* and *B. licheniformis* [10, 27-30], but the polymer obtained is mainly formed of D-units. As γ-PGA produced by *B. licheniformis* is an exocellular polymer that is freely excreted into the fermentation broth, its recovery is relatively easy. The broth was first centrifuged to remove the cells (9000 rpm, 30 min). The γ-PGA in the supernatant was then precipitated using cold ethanol (3:1 volume ratio) to concentrate and partially purify the polymer. The resulting crude γ-PGA was then redisolved in 100 volumes of distilled water and filtered to remove any remaining solids. Next, the PGA solution was dialyzed (to remove salts and low-molecular weight impurities) and concentrated with a S-Minitan equipment and Prep/Scale-TFF cartridge (Millipore). The polymer was then rotational evaporated to finally obtain a film, which was then redisolved to a final volume of 10 ml in distilled water. This PGA solution was acidified with 12M HCl and precipitated with 100 ml of cold 1-propanol. The polymer was finally washed with ethanol and air dried.

Table 2. Composition of modified Sauton's medium

Content [a]	Concentration (g/L)
L-glutamic acid	20
Glycerol	80
Citric acid	12
NH_4Cl	7
$MgSO_4$	0.5
$FeCl_3 \cdot 6H_2O$	0.04
K_2HPO_4	0.5
pH is adjusted to 7.4 with NaOH	

[a] 15 g/L of agar are added to the medium to prepare agar plates.

Table 3. Composition of the medium E for production of γ-L-PGA by *B. licheniformis* NCIMB11709 (adapted from Leonard et al., 1958 [25])

Content	Concentration (g/L)
L-glutamic acid	20
Glycerol Citric acid	80
Citric acid	12
NH_4Cl	7
$MgSO_4{\cdot}7H_2O$	0.5 (2.03×10^{-3} M)
$FeCl_3$ $6H_2O$	0.04 (1.44×10^{-4} M)
K_2HPO_4	0.5 (2.88×10^{-3} M)
$CaCl_3$ $2H_2O$	0.15 (1.02×10^{-3} M)
$MnSO_4$ H_2O [a]	0
pH is adjusted to 6.5 with NaOH	

[a] The concentration of Mn^{+2} can be varied from 1.54×10^{-7} M to 2.46×10^{-3} M to give the desired enantiomeric content in γ- PGA. In this study we used the medium without $MnSO_4$, to obtain polymer with maximum L content.

The number- and weight-average molecular weights (Mn and Mw, respectively) of the synthesized polymer, chemical structure, and stereochemical composition were performed as described by Pérez-Camero et al. [26]. Mn and Mw were determined by gel permeation chromatography (GPC). Aliquots of synthesized PGA were filtered through a 0.45 μm cellulose acetate syringe filter unit (Millipore) to remove the cells, and then directly injected into the GPC column using a Perkin-Elmer LC 2500 system equipped with two Ultrahydrogel™ linear columns (Waters) and a UV detector (Applied

Biosystems 785 A) operating at 220nm [26]. Chemical structure of the γ-PGA was evaluated by NMR (Figure 2).

^{1}H and ^{13}C NMR spectra were recorded on a Bruker AMX-300 spectrometer operating at 300.13 and 70.48 MHz, respectively [26]. The stereochemical composition of γ-PGA was determined by HPLC as described by Pérez-Camero et al. [26].

The γ-PGA obtained in this study had a L-glutamic acid content of 70% (γ-70% L-PGA), higher than that obtained by biosynthesis previously reported by other authors with *B. licheniformis* [8] and similar to that obtained by Pérez-Camero et al. [26]. The final yield of the polymer was 90 mg/L, similar to that obtained by these authors when no $MnSO_4$ was added to the fermentation medium. Low volumetric yields are obtained in cultures with no or very low $MnSO_4$ concentrations, since Mn^{2+} affects cell viability [7].

The Mn and Mw of poly (γ-L-glutamic acid) are shown in Table 4. Taking into account that Mn^{2+} was not present in the fermentation medium, high molecular weight for γ-70%L-PGA has been obtained in this work, which is relatively higher than that previously described [26]. Culture conditions, like incubation time, might have determined the final molecular weight of the polymer.

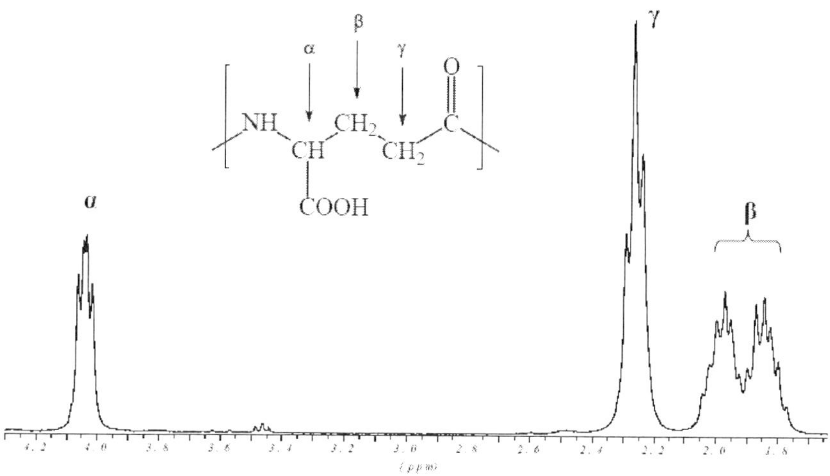

Figure 2. ^{1}H-NMR spectrum of pure γ-D-PGA obtained by fermentation. Solvent was D2O. Doublet assigned to β position is due to effect of neighbor chiral carbon. Signals of hydrogen atoms in COOH and NH are not present because they interchange rapidly with D2O.

Table 4. Molecular weight of the polymers before and after biodegradation assays

Polymer	Origin	% D:L	Mn		Mw	
			Initial	Final	Initial	Final
γ-70%L-PGA	biosynthesis	30:70	193000	61000	986000	228000
γ-77%D-PGA	biosynthesis	23:77	462100	81700	1954000	272000
γ-D,L-PGA	Meiji Co	59:41	147000	912000	388000	288000
α-L-PGA	Sigma-Aldrich	0:100	53000	15000	154000	33000
α-D-PGA	Sigma-Aldrich	100:0	36000	34000	83000	81000

Biodegradation

The ability of *B. licheniformis* NCIMB 11709 to degrade the different enantiomeric forms of PGA, as well as its α and γ forms (γ-L-PGA, γ-D-PGA, γ-D, L-PGA, α-L-PGA and α-D-PGA) was investigated. For this purpose, a basal medium supplemented with the corresponding polymer solution at a final concentration of 0.25% was used. The composition of the basal medium is shown in Table 5. The pH was adjusted to 7.0. For each culture, a 100 ml Erlenmeyer flask containing 30 ml of the sterilized basal medium and 0.25% (w/v) of filter sterilized polymer solution was inoculated with 300 μl of bacteria and was incubated at 37°C with shaking. The inocula was obtained from an overnight culture at 37°C in tryptone soy broth (TSB) of *B. licheniformis* NCIMB 11709; the cells were washed twice with 6 ml of basal medium and finally resuspended in 5 ml of basal medium before use. Degradation assays were carried out for over 260 h. As control, cultures without polymer or bacteria were also performed. Bacterial growth in culture medium, polymers concentration, and their average molecular weights were routinely measured at various incubation times. Bacterial growth was monitored by the level of absorbance at 660 nm using a Pharmacia Biotech Novaspec II UV spectrophotometer. Mn and Mw were measured removing the cells by filtration [17] and carrying out GPC analyses using the same above described chromatographic conditions for the determination of the molecular weight of poly (γ-L-PGA). PGA concentration was also measured by GPC.

Control runs for hydrolytic degradation, where polymer samples (γ-L-PGA, γ-D-PGA, γ-D, L-PGA, α-L-PGA and α-D-PGA) were incubated over identical exposure conditions in the absence of bacteria, showed negligible molecular weight loss. Also, no bacteria could grow in the basal medium without polymer (Figures 3 and 4).

Table 5. Composition of the basal medium for biodegradation assays

Content	Concentration (g/L)
KH_2PO_4	0.7
K_2HPO_4	0.7
$MgSO_47\ H_2O$	0.7
NH_4NO_3	1.0
NaCl	0.005
$FeSO_4{\cdot}7\ H_2O$	0.002
$ZnSO_4{\cdot}7\ H_2O$	0.002
$MnSO_4{\cdot}1\ H_2O$	0.001
pH is adjusted to 7.0 with NaOH	

B. licheniformis could grow using poly-γ-glutamic acid as carbon and nitrogen source, regardless of its D- or L- composition (Figure 3). The biodegradation studies of PGA have usually been performed with the polymer composed mainly of D- units, as it is the form of the polymer obtained when culturing the bacteria in the optimal conditions to achieve high yield of PGA. Here, we analyzed the three enantiomeric forms of the polymer (D-, L,- and racemic), and a similar behaviour of the bacteria was observed.

The three polymers were degraded by *B. licheniformis* to a similar final molecular weight, ranging from 227 kDa to 288 kDa. However, the higher decrease of molecular weight was observed in γ-D and γ-L-PGA (Table 4). Since *B. licheniformis* NCIMB 11709 is able to degrade both D- and L-PGA, as well as the racemic PGA, it might secrete different depolymerases which cleave the γ-glutamyl bonds between D- and D-, L- and L-, and D- and L-glutamic acid units. Several degrading γ-PGA depolymerases have been described, mainly from *B. subtilis*, but also from *B. licheniformis*. King et al. [18] characterized a γ-PGA depolymerase from *B. licheniformis* ATCC 9945a, which was tightly, but not covalently, associated with γ-PGA and showed an *endo*-D type activity. The enzyme was activated by Zn^{2+} and Ca^{2+} ions and was more active at high pH (pH= 8-10). Based on the studies from King et al. [18] with γ-D-PGA and the γ-PGA *endo*-D-depolymerase, Ashiuchi and Misono [27] suggested the biodegradation mechanism of γ-PGA: highly elongated γ-PGA is first digested by γ-PGA *endo*-L- and *endo*-D-depolymerases to yield γ-PGA with low molecular masses, which are further hydrolyzed to oligo-γ-glutamic acids or the amino acid monomer by γ-PGA *exo*-depolymerase. The results from our study strongly support this hypothesis, since all γ-PGA were degraded by *B. licheniformis* NCIMB 11709, independently of the L- or D- contents of the polymer.

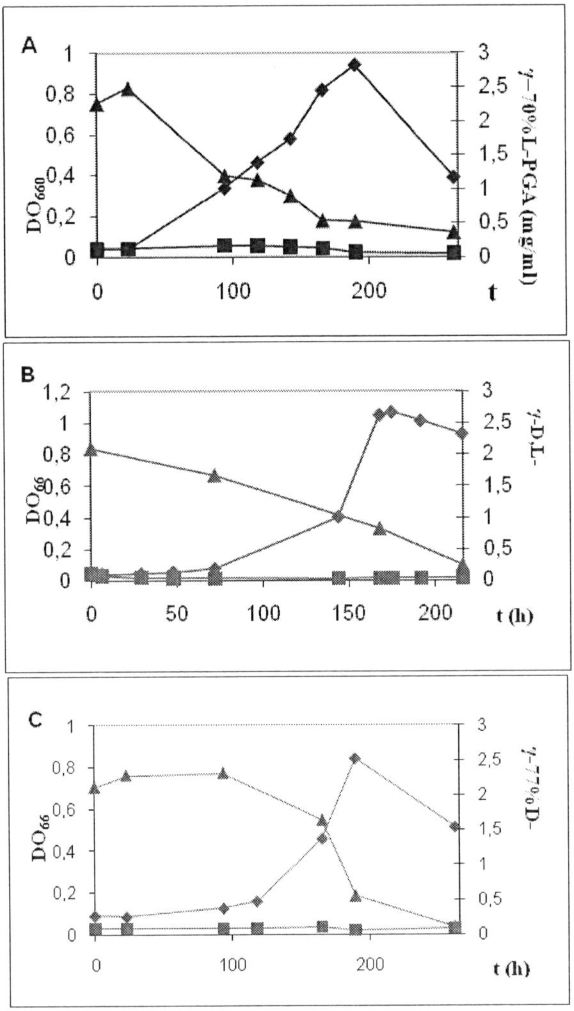

Figure 3. Biodegradation of γ-PGAby *B. licheniformis* NCIMB 11709. A: biodegradation of γ–L-PGA; B: biodegradation of γ-D, L-PGA; C: biodegradation of γ-D-PGA. Polymer concentration is indicated by triangles (▲); growth in minimal medium is indicated by squares (■); growth in minimal medium with γ-PGA is indicated by diamonds (♦).

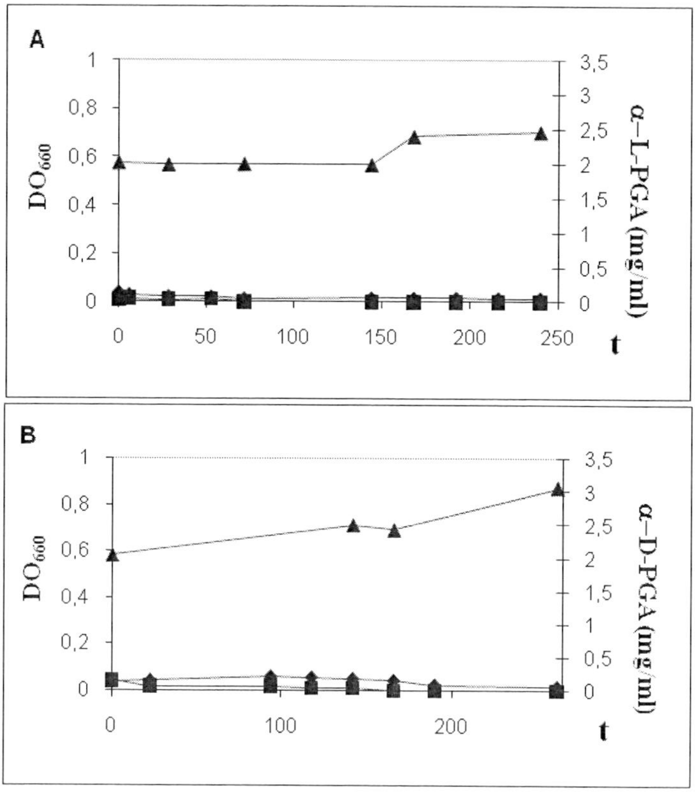

Figure 4. Biodegradation of α-PGA by *B. licheniformis* NCIMB 11709. A: biodegradation of α–L-PGA; B: biodegradation of α–D-PGA. Polymer concentration is indicated by triangles (▲); growth in minimal medium is indicated by squares (■); growth in minimal medium with α–L-PGA or α–D-PGA is indicated by diamonds (♦).

Lately, several studies have been published regarding the isolation and characterization of PGA-degrading enzymes from *Bacillus* species, with different stereospecificities and *exo*- or *endo*-activities. Kimura et al. [31] characterized a γ-glutamyltransferase (GGT) from *B. subtilis* NAFM5, involved in the degradation of capsular poly (γ-glutamic acid), which contains D- and L-glutamate. The enzyme was synthesized during the stationary phase and the capsule γ-PGA was first internally degraded by an *endo*-type of γ-PGA hydrolase into 1×10^5 Da intermediates, then externally into glutamates via GGT. This GGT had a powerful *exo*-γ-glutamyl hydrolase activity. GGTs from *B. subtilis* (natto) seem functioning as γ-PGA *exo*-depolymerases. These enzymes hydrolyze more effectively the fragmentized γ-PGA than the highly elongated γ-PGAs [32]. Ashiuchi et al. [19] described the production and purification of an *endo*-D-depolymerase from *B. subtilis* (*chungkookjang*). It was produced by cloning the *pgdS* gene in *E. coli*. The enzyme digested PGAs from *B. subtilis* (D-glutamate content, 70%) and from *B. megaterium* (D-glutamate content, 30%), but not from *Natrialba aegyptiaca*, which consists only of L-glutamate. Further studies from Ashiuchi et al. [20] described a novel *endo*-peptidase that cleaves the γ-amide linkage between two D-glutamate residues in DL-PGA. Another hydrolase, the YwtD hydrolase cloned from *B. subtilis* IFO 16449, was shown to be a γ-DL-glutamyl hydrolase, cleaving only the γ-glutamyl bond between D- and L-glutamic acids of PGA. The enzyme acts in an *endo*-type manner [33]. This hydrolase has recently been further analyzed and it seems to act in a different way: it cleaves the γ-glutamyl bond between two consecutive D-glutamic acid residues, recognizing adjacent L-glutamic acid toward the N-terminal region of PGA [22]. Also, a non-PGA producer, *Bacillus* sp. CMU29, has been isolated from a natto-like product that produces an *endo*-type PGA-degrading enzyme [21]. This γ-glutamyl hydrolase cleaves the γ-glutamyl linkage between two contiguous L-glutamic acids in polymer chain of PGA. On the other hand, the PGA-depolymerase from *B. anthracis* catalyses the hydrolysis of the poly-D-glutamic acid capsule in an *exo*-peptidase manner [34]. Apart from all of these depolymerases from *Bacillus* species, it should be mentioned that the first characterized γ-PGA degrading exoenzyme was from *Flavobacterim polyglutamicum* [35]. The crude extract obtained from a culture of this bacterium completely degraded γ-L-PGA to free L-glutamic acid. Finally, a γ-PGA hydrolase from a bacteriophage that infects *B. subtilis* host cells was described [36]. The enzyme might contribute to the phage infection of *B. subtilis* strains that produce γ-PGA by providing a better accessibility of phage receptors on the cell surface.

On the other hand, information regarding the utilization and degradation by bacteria of α-PGA, both the D- and L- forms, is scarce. No *Bacillus* species producing α-PGA have been described. Only α-L-PGA polymer has been found in the cell walls of pathogenic mycobacteria [37], and naturally-occurring α-D-PGA has not yet been identified. Therefore, in this study we have also explored the possible degradation and utilization of α-PGA by *B. licheniformis*. The α-PGAs (D- and L-) used in this study are of synthetic origin, and it should be pointed out that the molecular sizes of synthetic polymers are low compared with those of naturally-occurring γ-PGA (Table

4). Interestingly, *B. licheniformis* could not grow when the medium was supplemented with α-PGA, regardless of the D or L contents (Figure 4), indicating that α-PGA (carbon source) cannot be utilized by this bacterium for its growth. In the medium with α-D-PGA, there was no bacterial growth, no decrease in the molecular weight of the polymer (Table 4) and no significant decrease of polymer concentration. Furthermore, in the medium supplemented with α-L-PGA, although there was no bacterial growth and no decrease of polymer concentration, a decrease in the molecular weight was observed (Table 4).

Since there was no hydrolytic degradation of the polymer, this degradation was enzymatic. This indicates that in the culture supernatant there are one or perhaps more enzymes secreted by the bacterium that degrade the polymer, although the bacterium cannot utilize the degradation products. These enzymes are stereospecific, since this behaviour was not observed with the α-D-PGA. Early studies by Troy [38] also detected an *endo*-type activity in the culture supernatant of *B. licheniformis*, which showed no activity against α-D-glutamyl polymers.

More recent studies have described depolymerases from fungal or bacterial origin with no activity against α-PGA. Tanaka et al. [39] described an *endo*-depolymerase from the fungus *Myrothecium* sp. TM-4222 which acted on γ-DL-PGA of *B. subtilis* (*natto*), but did not show any activity on α-L-PGA or α-D-PGA. Also, the purified YwtD hydrolase from *B. subtilis* IFO 16449 (a γ-DL-glutamyl hydrolase) did not have any *in vitro* activity, neither on α-L-PGA nor on α-D-PGA [33].

Conclusion

Culture conditions used in this work have allowed us to produce γ-PGA with a high percentage of L units (γ-70%L-PGA). The assays of biodegradation reveal that *B. licheniformis* NCIMB 11709 cannot grow with α-PGA as C source, regardless of its enantiomeric form. This incapability of use by the bacteria can be related to the fact that there is no *Bacillus* species producing α–PGA. However, the bacteria can degrade all forms of γ-PGA, independently of the D or L contents. Results obtained here can be useful in future investigations to easily select the better form of the poly-glutamic acid according to their broad range of applications.

Acknowledgments

The authors are grateful for the financial support received from the Comisión Interministerial de Ciencia y Tecnología (MAT2002-04600-CO2-O2 and MAT2006-05979). We also wish to thank Dr. Kubota for kindly providing us the poly (γ-D, L-glutamic acid) used in this study.

References

[1] Candela T., Fouet A. Poly-gamma-glutamate in bacteria. *Mol. Microbiol.* 2006; 60: 1091-1098.

[2] Shih I. L., Van Y. T. The production of poly-(gamma-glutamic acid) from microorganisms and its various applications. *Biores. Technol.* 2001;79:207-225.

[3] Richard A., Margaritis A. Poly (glutamic acid) for biomedical applications. *Critical Rev Biotechnol.* 2001;21:219-232.

[4] Sung M.-H., Park C., Kim C.-J., Poo H., Soda K., Ashiuchi M. Natural and Edible Biopolymer Poly-γ-glutamic Acid: Synthesis, Production, and Applications. *Chem. Rec.* 2005;5:352-366.

[5] Kubota H., Nambu Y., Endo T. Alkaline hydrolysis of poly (γ–glutamic acid) produced by microorganisms. *J. Pol. Sci: Part A: Polymer Chemistry* 1996; 34:1347-1351.

[6] Giannos S., Gross A., Kaplan D., Mayer J. Poly (glutamic acid) produced by Bacterial Fermentation. In: *Novel Biodegradable Microbial Polymers*. Dordrecht: Kluwer Academic Publishers; 1990. p. 457-460.

[7] Cromwick A. M., Gross R. A. Effects of manganese (II) on *Bacillus licheniformis* ATCC9945A physiology and gamma-poly (glutamic acid) formation. *Int. J. Biol. Macromol.* 1995;17:259-267.

[8] Cromwick A. M., Birrer G. A., Gross R. A. Effects of pH and aeration of gamma-poly (glutamic acid) formation by *Bacillus licheniformis* in controlled batch fermentor cultures. *Biotechnol. Bioeng.* 1996;50:222-227.

[9] Ko Y. H., Gross R. A. Effects of glucose and glycerol on gamma-poly (glutamic acid) formation by *Bacillus licheniformis* ATCC 9945a. *Biotechnol. Bioeng* 1998;57:430-437.

[10] Yoon S. H., Do J. H., Lee S. Y., Chang H. N. Production of poly-gamma-glutamic acid by fed-batch culture of *Bacillus licheniformis*. *Biotechnol. Lett.* 2000;22:585-588.

[11] Morikawa M., Kagihiro S., Haruki M., Tanako K., Branda S., Kolter R., Kanaka S. Biofilm formation by *Bacillus subtilis* strain that produces gamma-polyglutamate. *Microbiology* 2006;152:2801-2807.

[12] Troy F A. Chemistry and biosynthesis of the poly (γ–d-glutamyl) capsule in *Bacillus licheniformis*. In: *Peptide Antibiotics-Biosynthesis and Functions*. New York: Walter de Gruyter; 1982. p. 49-83.

[13] Gardner J. M., Troy F. A. Chemistry and Biosynthesis of the Poly (γ–D–glutamyl) Capsule in *Bacillus licheniformis*. *J. Biol. Chem.* 1979;254:6262-6269.

[14] Kambourova M., Tagney M., Priest F. Regulation of polyglutamic acid synthesis by glutamate in *Bacillus licheniformis* and *Bacillus subtilis*. *Appl. Environ. Microbiol.* 2001;67:1004-1007.

[15] Ashiushi M., Kamei T., Baek D. H., Shin S. Y., Sung M.-H., Soda K., Misono H. Isolation of *Bacillus subtilis* (chungkookjang), a poly-γ-glutamate producer with high genetic competence. *Appl. Microbiol. Biotechnol.* 2001;57:764-769.

[16] Troy F. A. Chemistry and biosynthesis of the poly (γ–d-glutamyl) capsule in *Bacillus licheniformis*. I. Properties of the membrane-mediated biosynthesis reaction. *J. Biol. Chem.* 1973;248:305-315.

[17] Birrer G. A., Cromwick A. M., Gross R. A. Gamma-poly (glutamic acid) formation by *Bacillus licheniformis* 9945a: physiological and biochemical studies. *Int. J. Biol. Macromol.* 1994;16:265-275.

[18] King E. C., Blacker A. J., Bugg T. D. H. Enzymatic breakdown of poly-gamma-D-glutamic acid in *Bacillus licheniformis*: Identification of a polyglutamyl gamma-hydrolase enzyme. *Biomacromolecules* 2000;1:75-83.

[19] Ashiuchi M., Nakamura H., Yamamoto T., Kamei T., Soda K., Park C., Sung M.-H., Yagi T., Misono H. Poly depolymerase of *Bacillus subtilis*: production, simple purification and substrate selectivity. *J. Molec Catal B: Enzym*. 2003;23:249-255.

[20] Ashiuchi M., Nakamura H., Yamamoto M., Misono H. Novel poly-gamma-glutamate-processing enzyme catalyzing gamma-glutamyl-DD-amidohydrolysis. *J. Biosci. Bioeng.* 2006;102:60-65.

[21] Chunhachart O., Hanayama T., Hidesaki M., Tanimoto H., Tahara Y. Structure of the hydrolyzed product (F-2) released from gamma-

polyglutamic acid by gamma-glutamyl hydrolase YwtD of *Bacillus subtilis*. *Biosci. Biotechnol. Biochem*. 2006;70:2289-91.

[22] Chunhachart O., Itoh T., Sukchotiratana M., Tanimoto H., Tahara Y. Characterization of gamma-glutamyl hydrolase produced by Bacillus sp. Isolated from Thai Thua-nao. *Biosci. Biotechnol. Biochem.* 2006;70:2779-2782.

[23] Oppermann F. B., Pickartz S., Steinbuchel A. Biodegradation of polyamides. *Polym. Degrad. Stabil.* 1998;59:337-344.

[24] Thorne C. B., Gómez C. G., Noyes H. E., Housewright R. D. Production of glutamyl polypeptide by *Bacillus subtilis*. *J. Bacteriol.* 1954;68:307-315.

[25] Leonard C. G., Housewright R. D., Thorne C. B. Effects of some metallic ions on glutamic polypeptide synthesis by *Bacillus subtilis*. *J. Bacteriol.* 1958;76:499-503.

[26] Pérez-Camero G., Congregado F., Bou J. J., Muñoz-Guerra S. Biosynthesis and ultrasonic degradation of bacterial poly (gamma-glutamic acid). *Biotechnol. Bioeng.* 1999;63:110-115.

[27] Ashiuchi M., Misono H. Poly-γ-glutamic acid. In Fahnestock SR and Steinbüchel A (Ed.), *Biopolymers*, vol. 7. Weinheim: Wiley-VCH; 2002. p. 123–174.

[28] Richard A., Margaritis A. Optimization of cell growth and poly (glutamic acid) production in batch fermentation by *Bacillus subtilis*. *Biotechnol. Lett.* 2003;25:465-468.

[29] Chen X., Chen S., Sun M., Yu Z. High yield of poly- γ -glutamic acid from *Bacillus subtilis* by solid-state fermentation using swine manure as the basis of a solid substrate. *Biores. Technol.* 2005;96:1872–1879.

[30] Soliman N. A., Berekaa M. M., Abdel-Fattah Y. R. Polyglutamic acid (PGA) production by *Bacillus* sp. SAB-26: application of Plackett–Burman experimental design to evaluate culture requirements. *Appl. Microbiol. Biotechnol.* 2005;69: 259–267.

[31] Kimura K., Tran L. S. P., Uchida I., Itoh Y. Characterization of *Bacillus subtilis* gamma-glutamyltransferase and its involvement in the degradation of capsule poly-gamma-glutamate. *Microbiology-Sgm* 2004;150:4115–4123.

[32] Abe K., Ito Y., Ohmachi T., Asada Y. Purification and properties of two isozymes of γ-glutamyltranspeptidase from *Bacillus subtilis* TAM-4. *Biosci. Biotechnol. Biochem.* 1997;61:1621–1625.

[33] Suzuki T., Tahara Y. Characterization of the *Bacillus subtilis ywtD* Gene, Whose Product Is Involved in γ-Polyglutamic Acid Degradation. *J. Bacteriol.* 2003;185:2379-2382.

[34] Uchida I., Makino S., Sasakawa C., Yoshikawa M., Sugimoto C., Terakado N. Identification of a novel gene, dep, associated with depolymerization of the capsular polymer in *Bacillus anthracis*. *Mol. Microbiol.* 1993;9:487-496.

[35] Volcani B. E., Margalith P. A new species (*Flavobacterium polyglutamicum*) which hydrolyzes the gamma-L-glutamyl bond in polypeptides. *J. Bacteriol.* 1957;74:646–655.

[36] Kimura K., Itoh Y. Characterization of poly-c-glutamate hydrolase encoded by a bacteriophage genome: possible role in phage infection of *Bacillus subtilis* encapsulated with poly-c-glutamate. *Appl. Environ. Microbiol.* 2003;69:2491–2497.

[37] Wietzerbin J., Lederer F., Petit J. F. Structural study of the poly-L-glutamic acid of the cell wall of *Mycobacterium tuberculosis* var *hominis*, strain Brevannes. *Biochem. Biophys. Res. Commun.* 1975;62:246-52.

[38] Troy F. A. Chemistry and biosynthesis of the poly (γ–d-glutamyl) capsule in *Bacillus licheniformis*. II. Characterization and structural properties of the enzymatically synthesized polymer. *J. Biol. Chem.* 1973;248:316-324.

[39] Tanaka T., Hiruta O., Futamura T., Uotani K. Purification and characterization of poly (γ-glutamic acid) hydrolase from a filamentous fungus, *Myrothecium* sp. TM-4222. *Biosci. Biotechnol. Biochem.* 1993;57:2148-2153.

[40] Hanby W. E., Rydon H. N. The capsular substance of *Bacillus anthracis*. *Biochem. J.* 1946;40: 297–309.

[41] Schneerson R., Kubler-Kielb J., Liu T. Y., Dai Z. D., Leppla S. H., Yergey A., Backlund P., Shiloach J., Majadly F., Robbins J. B. Poly (gamma-D-glutamic acid) protein conjugates induce IgG antibodies in mice to the capsule of *Bacillus anthracis*: a potential addition to the anthrax vaccine. *Proc. Natl. Acad. Sci. USA* 2003;100:8945-50.

[42] Aono R. Characterization of structural component of cell walls of alkalophilic strain of *Bacillus* sp. C-125. *Biochem. J.* 1987;245:467.

[43] Kandler O., König H., Wiegel J., Claus D. Occurrence of poly-γ-D-glutamic acid and poly-α-Lglutamine in the genera *Xanthobacter*, *Flexithrix*, *Sporosarcina* and *Planococcus*. *Syst. Appl. Microbiol.* 1983;4:34–41.

[44] Kocianova S., Vuong C., Yao Y., Voyich J. M., Fischer E. R., de Leo F. R, Otto M. Key role of poly-gamma-DL-glutamic acid in immune evasion and virulence of *Staphylococcus epidermidis*. *J. Clin. Invest* 2005;115: 688–694.

[45] Hezayen F., Rehm B. H. A., Tindall B. J., Steinbüchel A. Transfer of *Natrialba asiatica* B1T to *Natrialba taiwanensis* sp. nov. and description of *Natrialba aegyptiaca* sp. nov., a novel extremely halophilic, aerobic, non-pigmented member of the *Archaea* from Egypt that produces extracellular poly (glutamic acid). Polymer production by two newly isolated extremely halophilic archaea: application of a novel corrosion-resistant bioreactor. *Int. J. Syst. Evol. Microbiol.* 2001;51:1133-1142.

[46] Weber J. Poly (γ-glutamic acid)s are the major constituents of nematocysts in *Hydra* (*Hydrozoa, Cnidaria*). *J. Biol. Chem.* 1990;265:9664-9669.

Index

A

D

E

F

G

H

M

N

O

P

T

U

V

W

X

Y

Z